Electromagnetics

It's not Black Magic

Eldon Gordon PE

My Office

For Mary
56 years my bride
(they said it would not last)

Copyright

Paperback Edition

ISBN: 978-0-692-48732-7
LCCN: 2015911774

Barn Swallow Publications
4596 County Road 2629
Caddo Mills, TX
214-794-9306

PRINTED IN THE UNITED STATES OF AMERICA

Table of Contents

Introduction

This book is unique in two aspects. First, it is written in first person, with a small degree of humor. So, watch for the humor, sometimes it is a bit hard to recognize. I envision myself sitting down with you and discussing my 45+ years of experiences.[1] I especially want to discuss some of the unique and obscure properties of the electromagnetic (EM) wave phenomena. Mostly, these are things that you were not taught in your formal education. (Electromagnetic waves are also known as radio frequency, radio waves, radiation, etc; and I may use the terms interchangeably.)

Second, this is a simple book. I start with very basic concepts. If you already know these things, just breeze through them[2]. It will only take a few minutes. So, be patient, there is a method to my madness; more complexity follows. I will also present new equations that, to my knowledge, have not been previously published. But, even these new equations are characterized by simplicity.

However, you should have basic electrical knowledge. You should fully understand the following terms: Voltage, Current, Resistor, Capacitor, and Inductor. You should understand that radio communication ensues by an electromagnetic wave (EM) traveling between two locations.

---

[1] Sometime I use "I" and sometimes I use "we", whichever strikes my fancy at the moment.
[2] Or skip to block "Sources of Radiation".

My target audience is the radio frequency (RF) and/or antenna practitioner, who probably knows about Maxwell's equations, maybe even somewhat understands them, but may never use them directly in his/her[3] everyday work. Mr. Maxwell looks down on our toils while we, without even realizing it, indirectly use his miraculous equations.

This book will help you even if you have never heard of Maxwell. But, if that is the case, please take the time to find out about him. By the power of his mind he identified the equations describing how radio waves work. Lacking a means of testing, it remained for later generations to validate his equations. They were, and still are, correct. He was the father of our ability to communicate by radio. He should have a national holiday, but technical people seldom do. In fact, we are often denigrated because "technology is bad".

These pages are also for those who do not understand how radio (EM) works, but would like to know the basics. I cannot tell you how many times I have heard people say "It's 'black magic', no one really understands it". It is not "black magic" and it is the goal of this book to dispel that perception.

It is my hope that even those of you who know the subject well will be entertained and enlightened by this book. I present even well-known concepts in a more simplified way.

---

[3]This is the last time I will use "his/her"; henceforth, the pronouns mean "person".

Why do I believe such a book is needed? I have been an electrical engineer for 45+ years and specializing in EM and antennas for 25+ years (BS 1964 and MS 1986). I remember feeling inadequate for not understanding the subject as well as a thought I should. I was sure my peers were brighter engineers. Now I believe that we were all pretty much on the same plane. This book will make you feel better about yourself because now you will really understand what you already know.

Before jumping into the next section, let's define some terms:

Radio: A general term meaning transporting information, through space, from a transmitting site to a receiving site that recovers the information. Generally, the sites are separated by free space. Radio includes television, satellite television, cell phones, internet, hot spots, remote controls, etc.

Electromagnetics: Is an energy field that enables the propagation of energy between a transmitting site and a receiving site. The EM field is composed of the electric field (voltage) and the magnetic field (current). Both fields must be present for energy to propagate. This is real energy. Intense energy will burn you. There is a joke (hopefully) that if you smell bacon, get out of the radiation area. More than one traffic cop has lost his ability to have children because he laid his radar gun in his lap.

Radiation: Is an EM propagation of energy emanating from an energy source. I will sometimes use the terms interchangeably. In this book, radiation always emanates from an antenna.

Antenna: A device that launches the energy from the transmitter into free space. Another antenna (maybe of different physical form) literally sucks some of the energy out of the surrounding free space and sends it to the receiver. An antenna is bi-directional; i.e., transmits and receives at the same relative way; i.e., they are reciprocal.

Transmitter: In this book, a piece of equipment that produces the source of EM energy and applies it to an antenna.

Receiver: In this book, the piece of equipment that receives a very, very small portion of the source energy from an antenna and converts it to good use.

Magic

I will make references to various EM effects as being "magical"[4]. Some of these magical things may be supported deep in the bowels of Maxwell's equations, but some are not. Accept them as truth, life will be easier.

Backbone Parameters

Here are some parameters that are the backbone of EM equations:

Speed of light (c) is how fast an EM wave travels in free space. In any other medium, including air, the EM wave travels slower.

Wavelength "λ" is the ratio of the speed of EM radiation (light) to the frequency of oscillation. Wavelength (λ) appears in most EM equations. It has the dimension of length. Most EM equations require that the length be in meters. The relative wavelength size of an object, i.e.; an antenna, is extremely important. There are two worlds; one in which the objects are smaller than one wavelength and one where the objects are larger than one wavelength.

Pi (π) is the ratio of the circumference of a circle to its diameter. Also, 2π is the ratio of the circumference of a circle to its radius. These values appear in almost every EM equation.

The "e" is a number 2.718281 (to 6 decimal places). Like π, it goes on forever. Unlike π, "e" is not related to a physical constant. Also, unlike

---

[4] The term "magical" is not used here in the as sense as "black magic" in the book title. For example, it is magical that when water freezes, it becomes less dense whereas most other substances making the same transition do not; making life on earth possible.

π, “e” is used in a myriad of other fields of mathematical applications. It is a toss-up whether π or “e” is the greater mathematical miracle. I will use both extensively in the remainder of this book.

The “j” is the square root of minus one (√-1). Since the square root of a negative number is impossible, it is called the imaginary. (In other scientific fields, the imaginary is designated as “i”.) The miracle here is that an artificially, made up, parameter has such a huge impact in mathematics. In EM “j” is used to identify the phase of the waveform.[5]

Infinity (∞) comes from the concept that no matter how large a number is, you can always add more. So, does infinity really exist? Maybe not in the real world, but it is very important to the mathematics of EM. Along with infinity, zero is a very important number. Do you realize that in relatively recent history people did not even know there was such a number? I believe that schools should teach counting as 0, 1, 2, 3.... not 1, 2, 3.....

Impedance: This entry is not a parameter, but it is extremely important to the study and understanding of EM. Impedance has, as the name suggests, the property to impede, to limit, to resist. There are two types: first is the impedance of physical devices; i.e., resistor, capacitor and inductor. Second, is the effect of impeding the progress of an EM signal. This second type is more correctly expressed as “characteristic impedance”, but by common usage, the term

[5] In electricity i or I has been used to designate current. The j is used to avoid that confusion.

"characteristic" is omitted. There is much more on this later.

Important note:
Characteristic impedance and wavelength are the essence of the electromagnetic effect.

Energy, Power and Work

Energy is a phenomenon available to generate power. Power is the act of performing work or generating heat. In physics, "work" is defined as moving a mass against a resistance. You can attempt to lift a barbell until you are exhausted, but if you did not move the barbell, you did not perform "work", but you did use energy by generating heat in your body. (You worked up a sweat, didn't you?). Examples in EM include the power supplied to a TV receiver to generate the picture (light) and the sound; or in the heat definition where it is usually called "loss"; i.e., wasted energy.

The Objective

The objective is the transfer of information from one point to another point through an intervening space using the space as the medium for the transfer. The transfer of information requires the movement of energy between the two points. In this book we study only the bottom rows of the following figure; i.e., the items in the boxes, the antennas and the intervening space between the antennas.

The System

Energy is transposed from here to here. Along the way much of the initial energy is converted to work by the heating produced at each stage of the process. The only useful work is it the light from the screen. (Not including any heating to produce the light.) And the movement of the speaker to produce the audible sound. (Not including the heating of mechanism driving the speaker.)

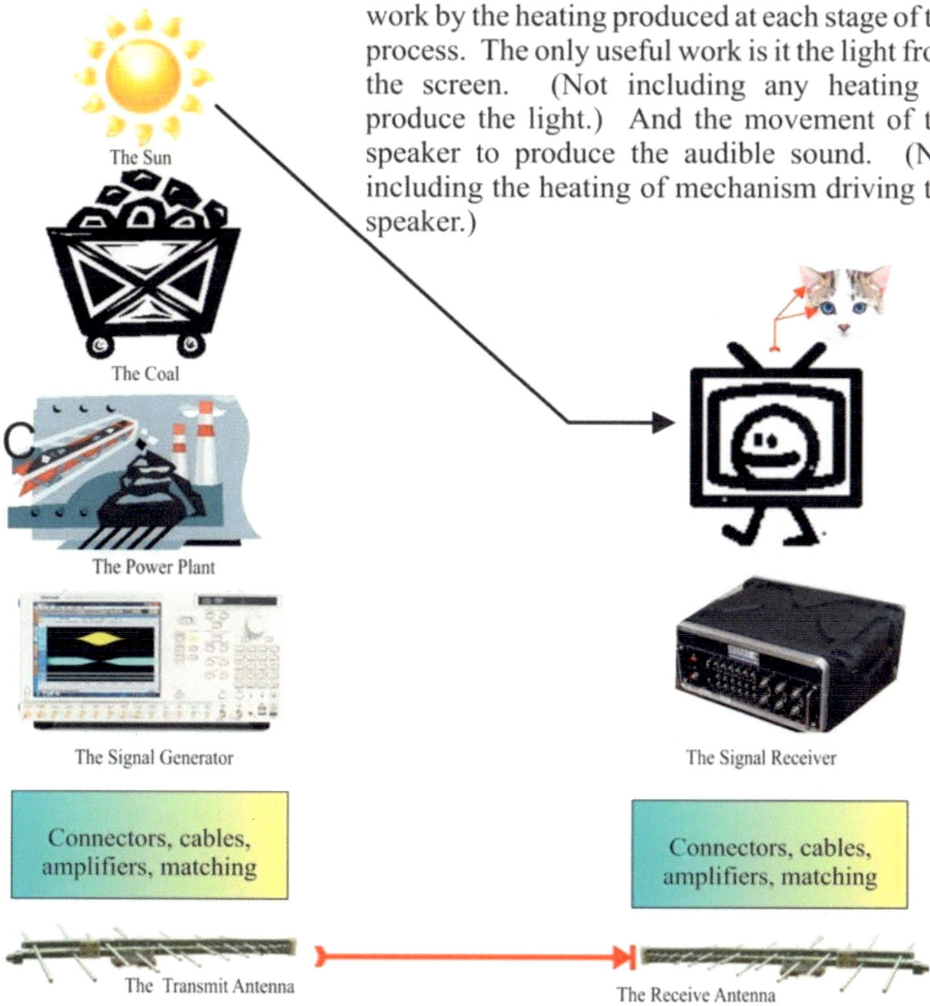

This path can be very short or very long, millions of miles. Example: After more than 30 years, NASA received the last signal from the venerable Pioneer 10 spacecraft at the Earth receiver on 23 January 2003. There is no heating loss in this path. The signal get weaker because it is spreading out. This applies to the path only, the antennas produce heat as the signal passes through them.

Defining the Backbone Parameters

Ok, I warned you this was going to be simple.

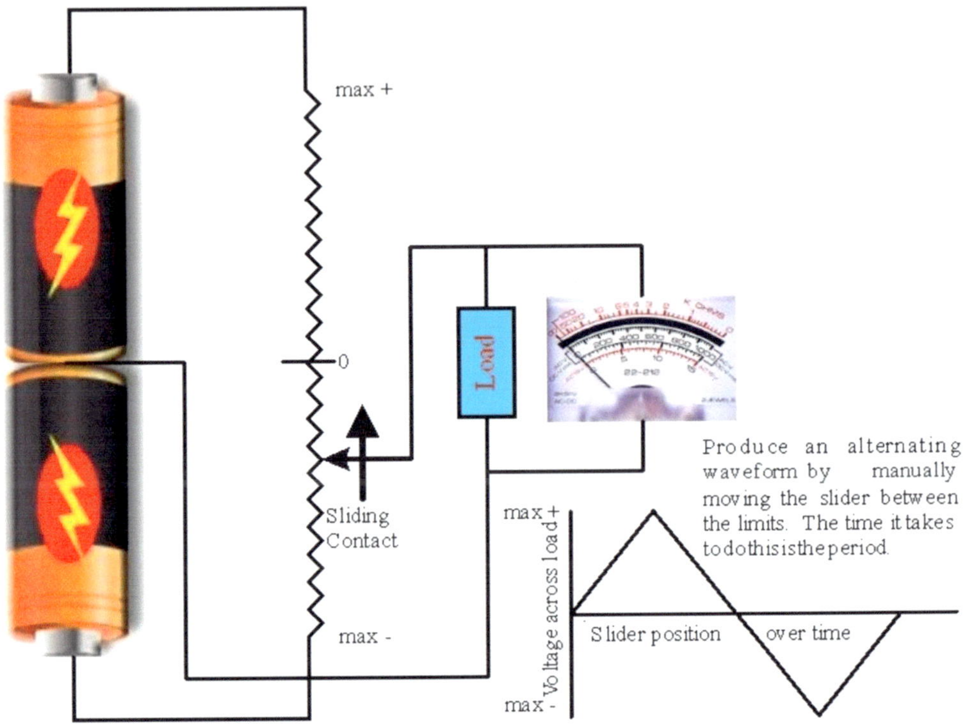

The first step is to accept the miracle that EM radiation travels at about 186,000 miles per second, both in space and a piece of wire. This diagram is used to define the wavelength, λ. Envision two batteries connected back-to-back with wires connected from the terminals and a common (ground) between them. A sliding control goes from zero volts to max (+) volts then back through zero to max (-) volts, then back to zero. If this interval from zero to zero to zero is repeated over and over in precisely the same way it is called the period and the dimension is time.

Remember that the wavelength (λ) is the speed of light divided by the frequency[6] of the wave.

Here is a very interesting fact. Both the speed of light and the value of pi are constants with an exact value. It is only by the accidental choice of our numbering system being base 10 that π is irrational. If humans had discovered π before his ten digits, we could have had a numbering system with base π. The speed of light is a universal constant. The value was made rational at 299,792,458 meters per second by adjusting the length of a meter. At the 1983 *Conference Generale des Poids et Mesures*, a meter was re-defined as the length that light traveled (in vacuum) in 1/299,792,458 of a second. With the advancements in accuracy of time measurements, it is conceivable that length of a meter could change again, but the speed of light will not change!!

---

[6] Formerly this was called cycles/second (cps). Now it is called Hertz (Hz) in honor of Heinrich Hertz.

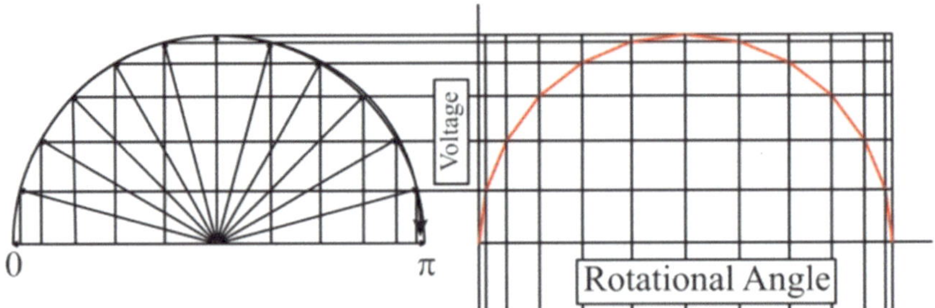

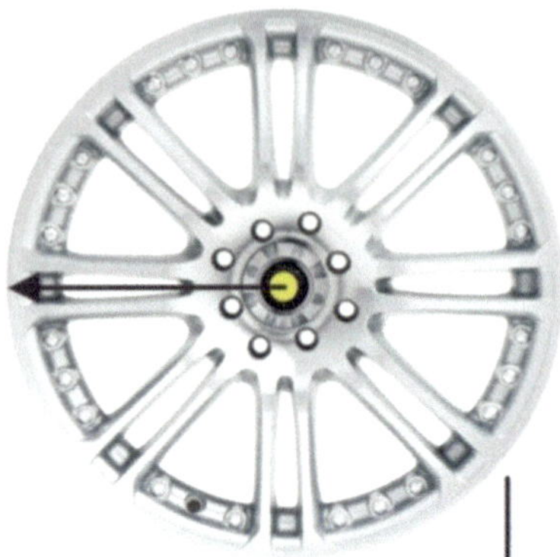

Consider that a spoke on the outer rim of the wheel is moving the slider.

Each rotational angle assigned the same amount of time.

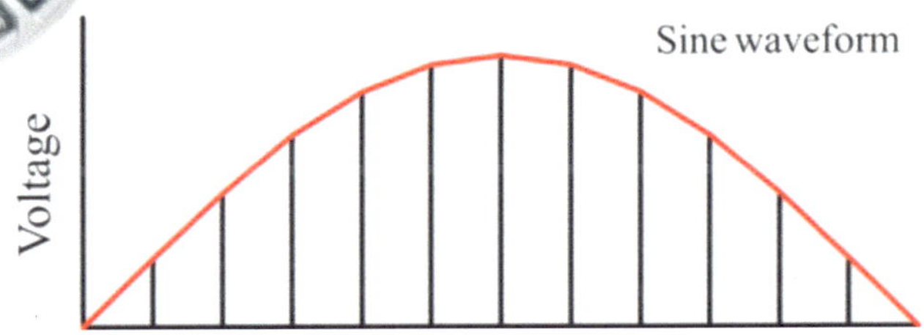

In real life electrical signals are not usually generated in a triangular waveform. Instead power plants use circular components to produce power. This circular rotation produces a sine wave form pattern as shown here. (Only one-half period is shown.) This provides the reason that π appears so much in EM equations.

A full rotation of the wheel is 2π and a full period of the signal is 2π. If the wheel makes one rotation in one second, the wavelength is 186,000 miles, more than half way to the moon. Household voltage is 60 cycles per second (Hz) with a wavelength of more than 3000 miles[7]; nearly across the United States. In this book I will mainly discuss frequencies above 10,000,000 cycles per second (10 MHz).

[7] But even the 3000 miles is significant because the massive interlocking energy grid in whole country must be kept in-phase so switching between grids is seamless.

More on Energy

Here is a good place for a reminder of the Law of Conservation of Energy. There is finite amount of energy in the universe. We are privileged to get most of ours from the sun. We get a very small amount from the other stars. Even "empty space" is not empty. The energy from "empty space" is called "background noise". Background noise was predicted earlier, but not proven until 1964. We encounter all of these in our work with electromagnetics.

Energy can be neither created nor destroyed; it can only be changed in form. Our objective is to move information (energy) from one location (transmitter) to another location (receiver) and convert it into work; i.e., the output of a radio.

Voltage, Current and Power

This is the last section of the really elementary stuff. After this I will get to real message of this book. In this first illustration, energy is supplied by a generator, moved through the wires to the new location where power is consumed as heat in the load (resistor).

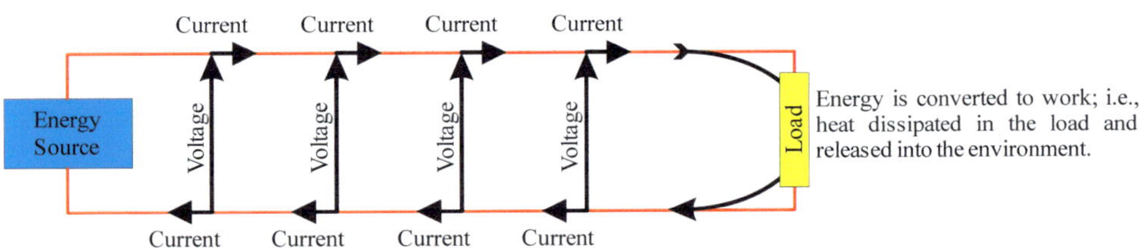

Energy is transported between locations. Along the way some of the energy is converted to heat because of resistance of the wire. Both voltage and current must be present for the transport to occur. If the energy source is a battery, all of the energy stored in the battery will be converted to heat in the load and the circuit will be dead.
If the energy source is an alternating current (ac) generator, the direction of the current changes direction within the period. The usual form of the ac c urrent is the sine wave, but it can be almost any form. The only constant is that every period is identical. Otherwise, the analysis described herein is not valid.

Then the next step is to move the energy without using wires. That is called electromagnetic (EM) propagation and it is one of the most wonderful miracles of our physical world.

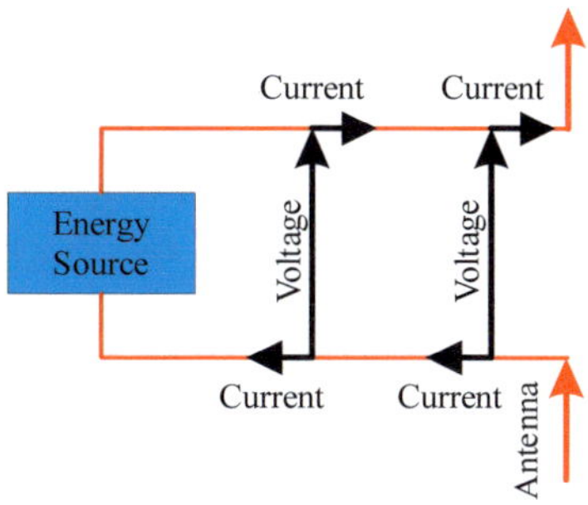

Electromagnetic (EM) propagation traveling at 186,000 miles per second. There is no loss of energy in the propagating wave.

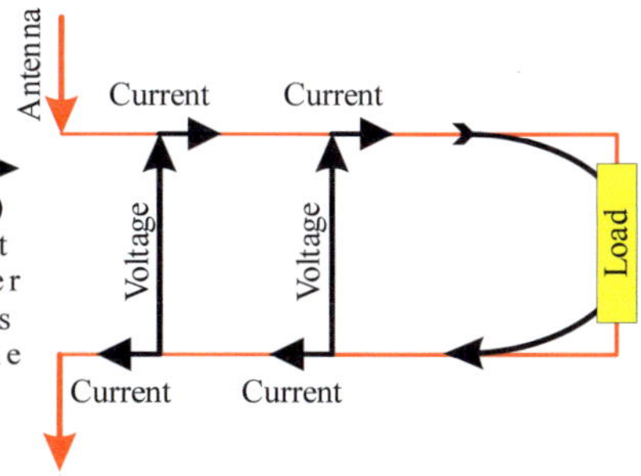

Miracle: The energy that reaches the antenna launches into free space and propagates away. Except for slight heating of the antenna, there is no loss of energy.

Miracle: A small portion of the EM energy is received by the antenna and transported to the load. The remaining very large portion of the propagating energy spreads out into free space. Some of it finds its way to outer space and continues forever. (The first television programs are now about 80 light years away and still going.) The remaining energy impinges on objects and is dissipated as heat.

In the intervening space:

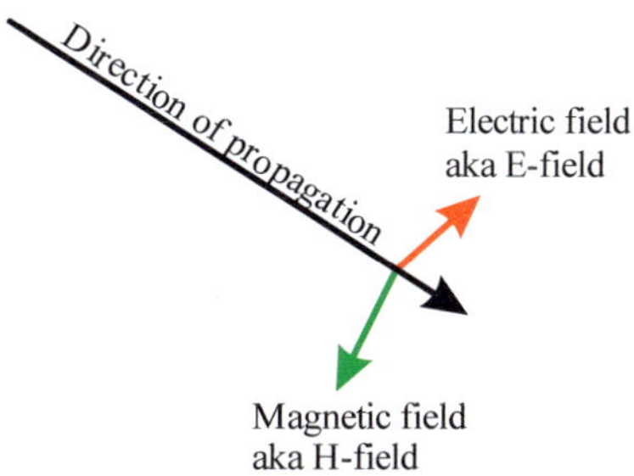

When the energy launches from the antenna, it forms into two fields that propagate away. The E-field corresponds to voltage in conventional circuits. The H-field corresponds to current. Both fields must be present for energy to be present. The two fields are perpendicular to each other and perpendicular to the direction of propagation. This phenomena is what this book is all about.

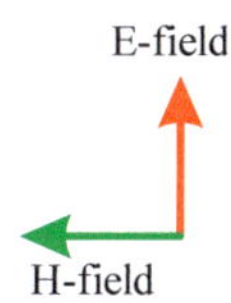

Source of Radiation

This is the most important paragraph in this book. All that preceeded and all that follows assumes the presence of EM radiation. The obvious delimina is: "How does radiation originate?" **When a force accelerates an electron, it radiates away some of the energy that caused it to accelerate**. That is a very powerful sentence. Think about that. Let it sink in. An accelerating electron losses some of the energy that caused it to accelerate, it detaches and radiates into space at 186,000 miles per second. The electron itself is not traveling, or accelerating, at anywhere near the speed of light. **Marvelous!!**

Electrons, Currents and Radiation

The definition of current is a movement of electrons. So, let's now investigate the mechanics of electron movement and how radiation results. In our work, radiation originates from conductors; i.e. copper transmission lines and metallic antennas. Not all transmission lines or antennas are metallic, but for this book I assume they are. Another term for transmission line is waveguide. There are many types of waveguides, but we will not get bogged down with that yet.

Current is a moving electron; or a moving electron is current. An electron moving at a constant velocity does not radiate energy. Each very small bit of a copper wire contains many millions of atoms. Each atom has a "free electron" that can easily be bumped from atom to atom. Before the switch is thrown, the millions of "free electrons" are moving from atom to atom, but since they are moving randomly in all directions, there is no net current. When an electric force is applied to an electron, it changes velocity (accelerates) to a new velocity in the direction of the applied force. The accelerating electron radiates some of its energy away. Conversely, if a copper wire receives radiation from an external source, the electrons in the copper are accelerated. The energy that is received travels at the speed of light down the wire.

Envision a copper wire filled with a cloud of widely spaced electrons; i.e., widely with respect to the size of an electron. (Remember that at the atomic level there is a lot of space between the nucleus and the electrons cloud.) When the switch is thrown ON, the nearest electrons are moved a very small distance, but they move (accelerate) very quickly. The first electrons emit radiation that is received by the next electrons in line. They accelerate and so on through the circuit. So, energy propagates down the wire at nearly the speed of light[8] until it reaches the bulb where it is transformed into work (heat and light). But the individual electrons moved only a very small distance. The velocity and acceleration of the electrons are much slower than the speed of light. (186,000 miles per second). If the length of the line in this circuit is 0.186 miles, the energy reaches the bulb 1/1,000,000 seconds[9] after throwing the switch. The net movement of the electrons is called the "drift velocity" and it is very small value. If the battery in this circuit is replaced by an alternating current generator, the net drift velocity is zero and the electrons move back and forth around a stationary point.

---

[8] Somewhere between 95% and 99% depending on whom you believe and whether the wire is bare or insulated.

[9] One microsecond. In EM, 1μs is a very long time.

Somewhat like rooted seaweed moving back and forth with the waves.

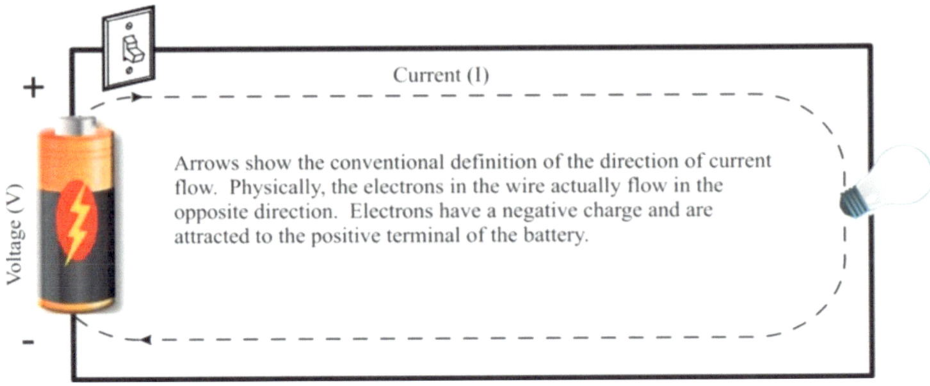

Consider this thought problem:
Remove the load and extend the length of the line to infinity. Let the copper be a perfect conductor (no loss). Throw the switch on. Energy propagates down the conductor pair at 186,000 miles per second. Energy is added to the line until the battery is discharged. Now we have an infinitely long line with a lump of energy traveling on it. No energy at the beginning and none at the end. The lump of energy travels forever down the infinite line and is never dissipated. So, the energy that was in the battery is still intact, totally contained and spread out over a length of the transmission line. If you could get ahead of the energy and place the light bulb on the end of the line you could light the bulb just as if the battery (now dead) did it. This has corollary to the first "I Love Lucy" show. There is about a half hour span of energy some 60+ light-years away traveling along on its merry way. Conservation of energy!! This energy is not lost, but it has definitely left this planet. But, it is not recoverable because its level is much, much lower than the free space background noise. That is why I think SETI is a waste!!

Quantifying Radiation

So, how much radiation energy is given up by an accelerating current? I did not know until I researched this book that there is an equation for that value. In 1897 J. J. Larmor gave us the equation for an accelerating charge. Charge is not a static number like volts. It is dynamic and is defined as 6.24×10^{18} electrons passing a given point in one second when the current level is one ampere. The unit of charge is coulomb.

P is the radiated power in watts.
e is a single electron charge 1.6×10^{-19} coulombs
N is the number of charges
a is the acceleration of the charges, m/s^2
ε_0 is the permittivity of free space
c is the speed of light, m/s

$$\mathbf{P} \equiv \frac{(\mathbf{N}\cdot\mathbf{e})^2\cdot\mathbf{a}^2}{6\pi\cdot\varepsilon_0\cdot\mathbf{c}^3}$$

I was just about to write that in all of my many years of experience I have never heard of this equation. Then I discovered this quote by Dan Schroeder, Department of Physics, Weber State University.

"I've never understood why the Larmor formula isn't even mentioned in most introductory textbooks. In fact I never saw it in all of my undergraduate education. Yet it's simple to interpret, and it has all sorts of interesting uses. More generally, the Larmor formula ties electromagnetic waves to their source, making it clear that **radiation requires not just motion, but also acceleration of charged particles.** Once you have the Larmor formula, why bother to cover Maxwell's equations?" Emphasis is mine. I will go Dan one further, I did not see it in my graduate work either.

All of the parameters of the Larmor equation are straight forward, except acceleration "a". The acceleration can be any value. It is restricted only by the fact that an electron cannot be accelerated past the speed of light; although, particle accelerators can come very close. On the internet I saw the question "How much of its energy does a charged object (pith ball) radiate away if it is dropped from a height and falls to the earth?"

But, I am more interested in microwave signals. The value of "a" is not at all obvious. So, I set up an example signal. In a length of copper line one foot long and 0.02 inches diameter what does "a" need to be to produce one watt of radiation at 1.0 GHz frequency? As stated earlier, the drift velocity of the electrons is very low, but the acceleration that moves the electron to its next position is very high. I worked it out that the peak acceleration is 5,000,000 m/s^2 to produce 1 watt and the peak velocity is 0.0008 m/s is needed to produce the acceleration. This is a plot of values over a single period.

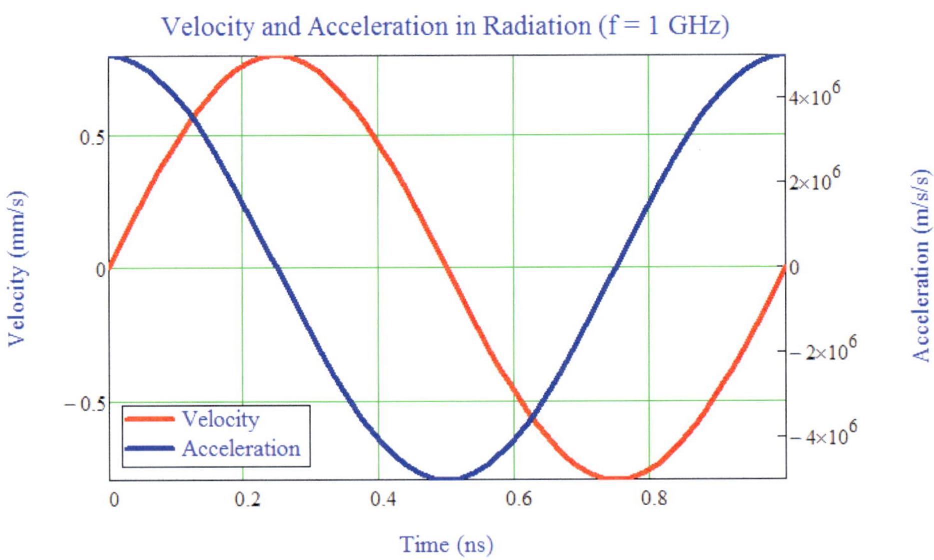

Analysis Methodology

Switching from current to voltage

The preceeding sections firmly established that electromagnetic (EM) radiation is created by current, more specifically, an accelerating current. The EM wave is composed of two components; the E-field and the H-field. They each contain all of the information needed to analyze EM performance. By convention, the E-field (voltage) is used in EM analysis. The H-field is disregarded because it is superflous. Any antenna can be analytically subdivided into small segments of current. The E-fields generated by each segment are added together to get the total antenna performance.

Here is good place to make a couple of points. First, EM analysis is primarily an exersise of adding sine waves together. Secondly, the EM waves have two components, time and distance, both oscillating as a sine wave. Look at the wave in two ways. First, hold time still and observe the sine wave amplitude as the distance from the source. Second, at a given point in space, observe the amplitude vary with time. In EM analysis, the first method is used. Time is usually not considered, just held constant.

The Mathematics of EM Analysis

As mentioned before, "e" is a truly miraculous number used in many applications of physics.[10] In

[10] From Wikipedia: With the possible exception of pi, "e" is the most important constant in mathematics since it appears in myriad mathematical contexts involving limits and derivatives. Among all exponential functions b^t, the one with base e has many special properties.

our case, it comes to rescue in the form of Euler's equation.[11]

Euler's Equation

$$e^{i\theta} = \cos(\theta) + i{*}\sin(\theta)$$

Euler's equation is shown here graphically Note the return of our old friend "i", the imaginary square root of minus one[12] in the lofty position as an exponent of "e". This is a very fortunate marriage of two miracle numbers. Any sum of a real number and an imaginary number is called a complex number; even if the real number is zero. This figure is the "complex number graph" that shows the graphical form of a complex

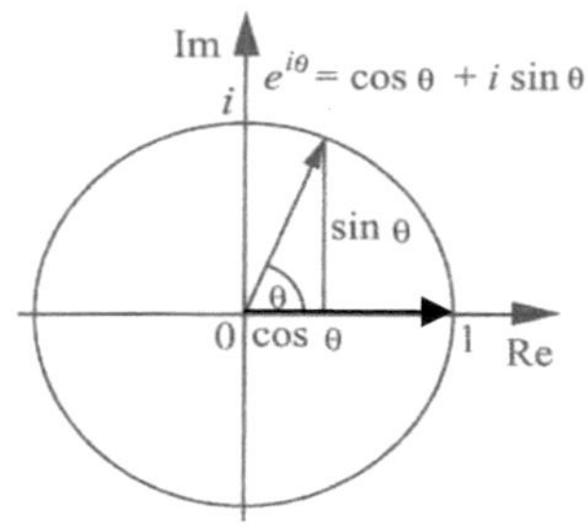

For any exponential function b^t, the slope of the tangent line is proportional to b^t. The constant of proportionality is 1 only when the base is the special number "e". This can be taken as the defining property of "e". As we shall see, this number "e" is not a whole number. In fact it's not even a fraction (such as 22/7 or 87/32). If you try to write it as a decimal, you find that the decimal expansion never ends and never repeats; instead it's an example of an irrational number.

[11] From Wikipedia: Euler's formula is ubiquitous in mathematics, physics, and engineering. The physicist Richard Feynman called the equation "our jewel" and "one of the most remarkable, almost astounding, formulas in all of mathematics."

[12] Remember that either "i" or "j" can be used as the imaginary (square root of minus 1). In EM we usually use "j", but are prepared for either.

number[13]. The real part is on the x-axis and the imaginary part is on the y-axis. Complex numbers are the backbone of EM analysis. Here's how:
This figure represents the way to add E-fields together. Our E-fields are sine waves with varying phase.

Sine Waves 60 Degress Out-of-Phase (Orange is the sum.)

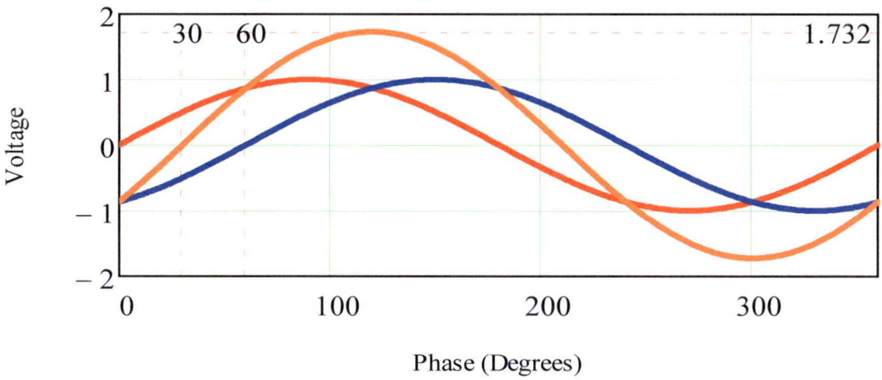

This example shows two sine waves phased 60 degrees apart (red and blue). Observe that the blue sine wave crosses zero 60 degrees later in time.[14] The two added together in time is the orange.) These sine waves can be represented as vectors as shown in the next figure.

---

[13] Graphic from Wikipedia

[14] Summing sine functions, regardless of how many there are or the phase differences always yield a single sine wave. BTW, cosine is just sine with a 90 degree phase difference.

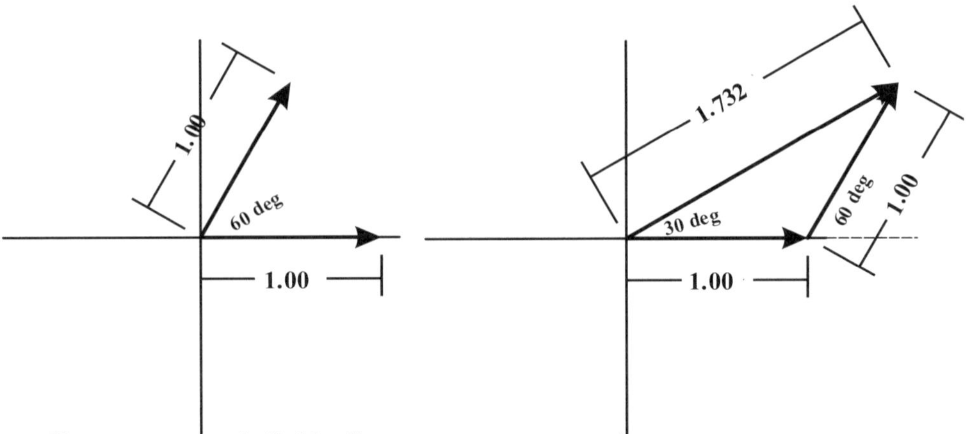

Sine wave vectors individually Sine wave vectors summed

Notice that both the time plots and the vector plots show the maximum value of the summed waveform is 1.732 and the phase of the summed sine wave is 30 degrees. This correspondence is the feature that allows us to use Euler's formula in the summation of sine waves.

Special Features of Sine Waves

Here are some important features of sine wave summation. Generally, we are summing E-field vectors in free space at a point of summation located very far away from the source of the sine waves. The bandwidth of a sine wave is zero. Only sine waves of the exact same frequency can be summed. **Sine waves of even very slightly different frequencies do not interact in any way and cannot be summed.** Signals with periods of other shapes, such as the triangular wave signal shown earlier, can be decomposed into a group of individual sine waves. There is more on this later. We almost always discuss the phase of sine wave as degrees. But, in reality, trigonometric functions operate on radians where 360 degrees corresponds to 2π radians. Some math software makes this conversion automatically. In others, you have to apply the conversion.

The truly wonderful thing about EM is that virtually all of the mathematics of **EM analysis is comprised of summations of individual sine waves.**

Scalars and Vectors

There are two types of units of measure, scalars and vectors. Scalars have a magnitude but no direction. Examples of scalars are 7 inches, 30 miles per hour. But 7 inches pointing 45 degrees above horizontal and 30 miles per hour northbound are vectors. The use of vectors occurs later in this book, but I want to include the characteristics of vectors in this tutorial section.

Any single vector can be subdivided into any number of "sub vectors" which when added together yield the single vector.

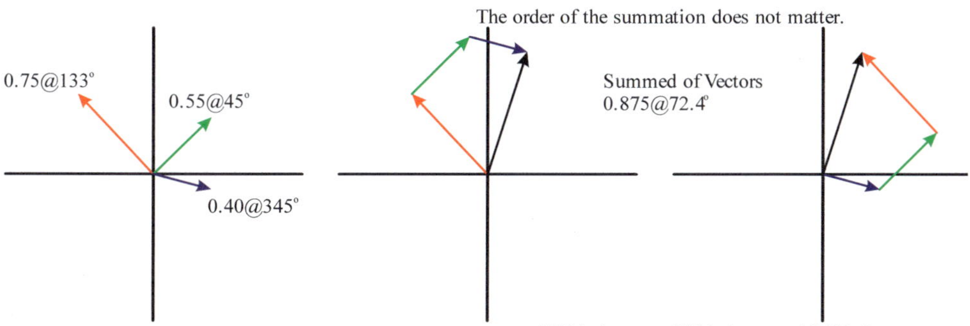

$$A \equiv 0.75 \cdot e^{j(133deg)} + 0.55 \cdot e^{j(45deg)} + 0.40 \cdot e^{j(345deg)}$$

$$A = 0.264 + 0.834i$$

$$|A| = 0.875 \qquad P \equiv \arg(A) \div deg = 72.4$$

To subtract a vector simply turn it around 180 degrees. See Appendix G for multiplying vectors.

Antenna Field Regions

Few aspects of EM engender as much confusion and misunderstanding as the concept of "far-field". I cannot count the number of times I have been asked, "How many wavelengths does it take to be in the far-field?" Well, that depends. There are generally three field regions.

Near Field

The near field is reactive; i.e., it acts like a inductor/capacitor. The radiated energy transfers from the antenna to nearby space then back to the antenna. Stored energy transferring back and forth like happens with an inductor or capacitor. The E and H fields are 90 degrees with respect to each other. Two inductors close to each other form a transformer which moves energy from one to the other. Likewise two antennas place closely together will couple energy from one to the other in the same way. The magnitude of the near field decreases with distance as the inverse of R^3 and R^2. (The "far-field" decreases as the inverse of R.) In the near field, there is no "radiation pattern".

Intermediate Field

In the intermediate region, the field is partly reactive and partly radiative. The radiation pattern is beginning to form, but it may change shape with distance from the source. I have searched the web and found at least four different definitions of the extent of the intermediate field; i.e., where far-field begins. The definition I like best comes from Rajeev Bansal published in *Applied Microwave & Wireless*, no date given. He gives attribution to Stutzman and Thiele, *Antenna Theory and Design*, 2nd edition, McGraw-Hill, 1998, page 53. His "rule of thumb" is to be in the far-field all of these conditions must be met:

$R > 2D^2/\lambda$

$R > 5D$

$R > 1.6\lambda$

D = maximum dimension of the antenna

λ = wavelength

The important thing to note is that two antennas will couple and provide an output signal no matter the distance between them. But you probably will not know the relative gains and patterns so you cannot calibrate, but you can make similarity comparisons among antennas. **It is not always necessary to be in the far-field.**

Far-field

For a large antenna (or array of antennas), observe two points, one at each edge. The EM wave emanates from each point as sphere; i.e., it travels in every direction from that point at the speed of light. Remember the previous exercise where we held the oscillation component constant and looked at the sine wave nature of the E-field as a function of distance. The distance that an EM wave travels can be described in portions of a wavelength (λ). At 1 GHz (1,000,000,000 cycles per second), it takes (1/1,000,000,000) seconds to travel one wavelength (11.8 inches) from the source point.

Let's look at far-field graphically.

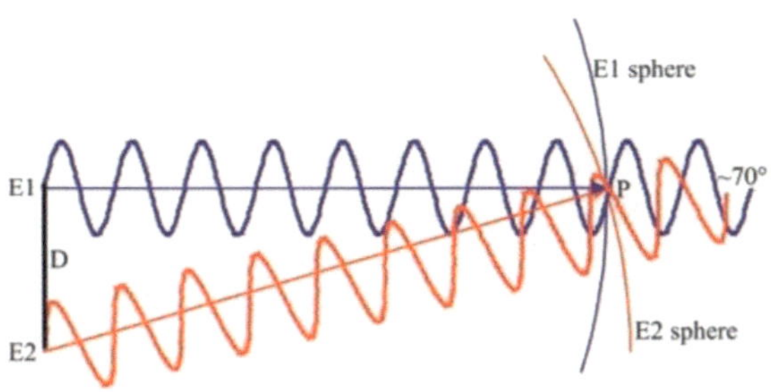

For this example

$$A = 1 \cdot e^{j \cdot (0 \deg)} + 1 \cdot e^{j \cdot (70 \deg)}$$

$$|A| = 1.638$$

$$Ph = \arg(A) \div \deg = 35$$

For conventional far-field

$$A = 1 \cdot e^{j \cdot (0 \deg)} + 1 \cdot e^{j \cdot (22.5 \deg)}$$

$$|A| = 1.962$$

$$Ph = \arg(A) \div \deg = 11.25$$

$$\lambda = 11.8 \text{in}$$

Select two points on the antenna (E1 and E2) and a point in space (P). The EM waves from E1 and E2, separated by D/2. E2 must travel farther than the wave from El to reach P. In this graphic, the difference is about 70 degrees. So, what is the total E-field at P from just these two waves? Answer: 1.638 volts with 35 degrees phase shift.

The most common definition of far-field is $2D^2/\lambda$ which corresponds to a phase difference of 22.5 degrees. In this case the summed voltage is 1.962 with a resulting phase of 11.25 degrees.

As the distance from the antenna to point P increases, the summed voltage approaches 2.0 and the phase differential approaches zero degrees. **This is the infinite far-field.**

Logarithms and dB

In EM the relative difference between power levels is huge. We solve this by defining the dB (decibel) by a logarithm. Let the ratio ($Power_2/Power_1$) = R. Then, dB = $10*\log_{10}(R)$. Where $\log_{10}(4356) = 3.639$ which is the number to which10 must be raised to equal 4356; i.e., $10^{3.639}$ = 4346. So, a power ratio of 4356 is 36.39 dB. The 10 is the base of the logarithm[15]. For a voltage ratio it is $20*\log_{10}(RV)$, but both voltages in the ratio must be related to the same impedance, otherwise the equation is meaningless.

It is common practice to set one milliwatt (1/1000 Watt) as a reference level. Then a power level is called xx dBm, meaning xx dB above (or below) one milliwatt.

Integrals

I know I promised no high level math, but I have to cover integrals. But the integral is nothing more than finding the area under a line or the volume under a surface. Example: consider a dining fly 10' x 10' at 7' above the ground. If the

[15] Logarithms can be to any base. In EM we use base 10.

dining fly is stretched taunt, the volume is 700 cubic feet (ft^3). Now let the dining fly assume a multi-curved shape. Taking the integral of the shape yields the volume. Let's take an example of the sine function raised to the 3rd power. Integration means we want to know the area under the plot of the function. You can get the value using the calculus definite integral. But, you can also get it by subdividing the phase-axis into small increments and extending a rectangle up to the plot value. Then sum all of the rectangle areas to yield the final value. This fortunate because much of the data we gather in EM analysis is sampled data.

We make extensive use of integration in EM analysis, design and validation. But, do not let that scare you off. Remember, it is just area under a curve or volume under a surface.

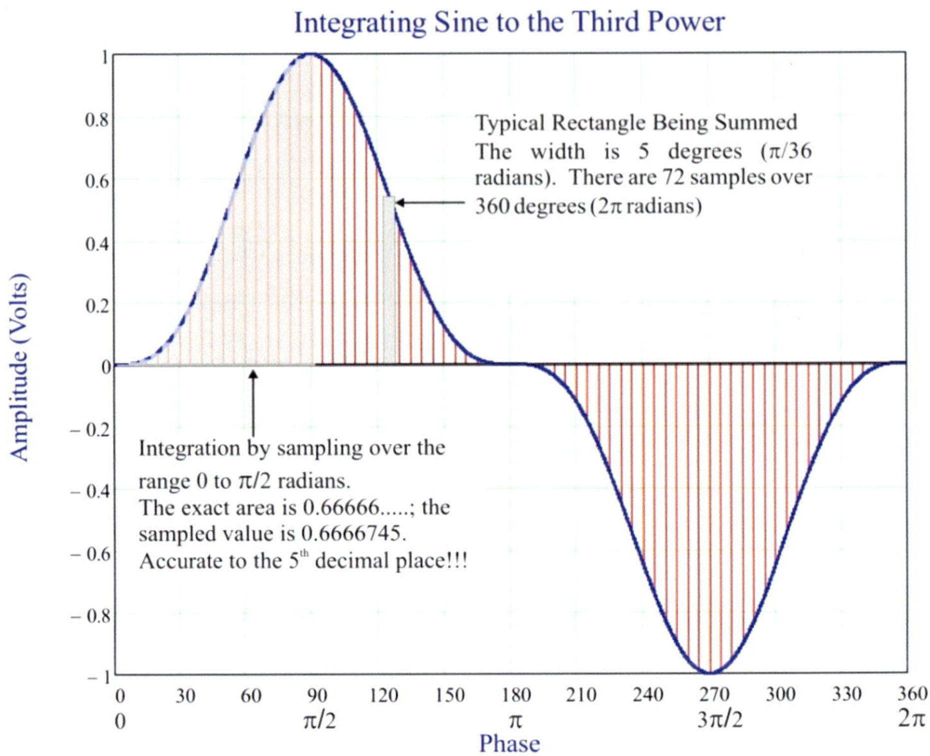
Integrating Sine to the Third Power
Typical Rectangle Being Summed
The width is 5 degrees (π/36 radians). There are 72 samples over 360 degrees (2π radians)
Integration by sampling over the range 0 to π/2 radians.
The exact area is 0.66666.....; the sampled value is 0.6666745.
Accurate to the 5th decimal place!!!
Amplitude (Volts)
Phase
1
0.8
0.6
0.4
0.2
0
– 0.2
– 0.4
– 0.6
– 0.8
– 1
0 30 60 90 120 150 180 210 240 270 300 330 360
0 π/2 π 3π/2 2π

Integration by Definite Integral of the Waveform

Over the range 0 to π/2

$$\int_0^{\frac{\pi}{2}} \sin(\theta)^3\, d\theta = 0.6666667$$

Over the range 0 to π

$$\int_0^{\pi} \sin(\theta)^3\, d\theta = 1.3333333$$

Over the range 0 to 3π/2

$$\int_0^{\frac{3\pi}{2}} \sin(\theta)^3\, d\theta = 0.6666667$$

Over the range 0 to 2π

$$\int_0^{2\pi} \sin(\theta)^3\, d\theta = 0$$

Integration by Sampling the Waveform

Number of Increments to be Summed

$$n \equiv 18$$

The Voltage Value for Each Increment

$$VV(x) \equiv \sin\left(\frac{x\cdot\pi}{n}\right)^3$$

The Width of Each Sample (Radians)

$$Wid \equiv \frac{2\pi}{72}$$

The Summation of the Samples

$$\sum_{x=0}^{n} [(VV(x))\cdot Wid] = 0.6666745$$

Antennas and Radiation

The purpose of an antenna is to launch an EM wave into free space or extract EM energy from free space. There are two elementary antennas.

First, let's deal with the universal standard. Imagine a huge, yet perfect, sphere. Inside the sphere is your transmitter, antenna, your house, your family, your coworkers and your pets.

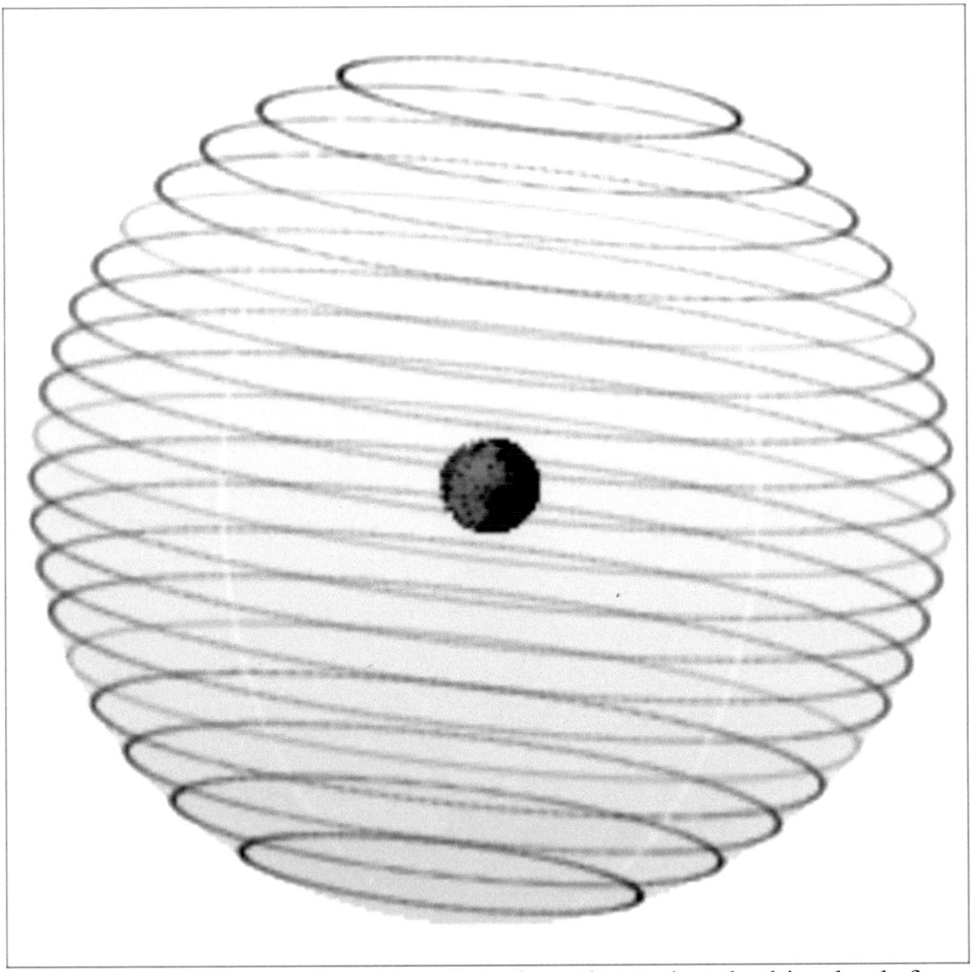

But this sphere is so large that, looking back from its edge, everything is infinitesimally small. Regardless of the size, let the radius equal 1.0. This is the radiation pattern of the fictitious omni-directional antenna. The directive gain[16] of any other antenna is usually related to this by the term dBi, meaning directive gain relative to isotropic.

Second, let's consider another impossible antenna; i.e., the infinitesimally small doublet (aka dipole).

---

[16] The term directive gain means the level of the antenna pattern relative to the isotropic sphere. Every direction from the antenna has a particular level, that is, greater than or less than the isotropic sphere. However, it is universally understood that the directive gain of an antenna is the maximum value, regardless of the relative direction to the antenna. Also, we seldom say "directive gain", we usually just say "gain". However the term "gain" alone must account for any losses in the antennas. Since very many antennas have very low loss, we can usually get away with this shortcut.

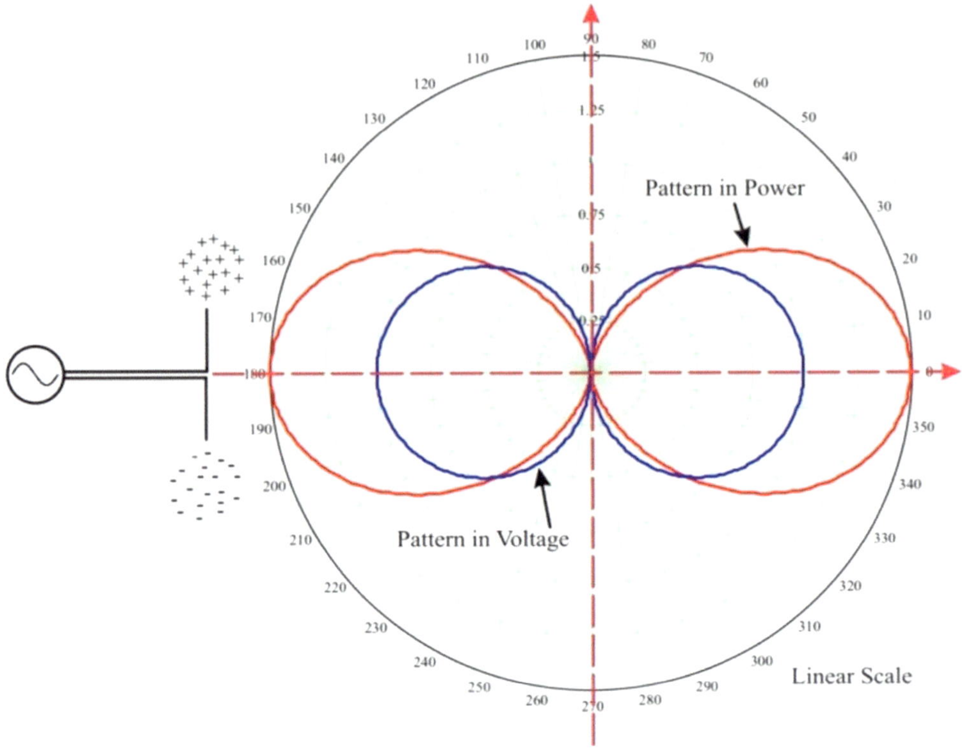

Let's remember that radiation occurs when an electron is accelerated from one point to another, meaning in straight line, the doublet. As previously stated, the fictional omni-directional antenna has a directive gain of 1.0 in every direction. So, it follows that any other antenna has a directional gain of greater than 1.0 in some direction and less than 1.0 in some other direction.

Now let's place the doublet in the center of the omni sphere and look at its radiation pattern. Note that the pattern in the direction parallel to the doublet is zero. That is because the charge clusters at each end of the doublet cancel each other. In the direction perpendicular to the doublet, the pattern has its maximum value of 1.5

power ratio (1.76 dBi).[17] Get the total pattern (3D) by rotating the plot around the vertical axis.

As a physical illustration, envision a spherical balloon. Now place a finger on opposite sides of the balloon. Press your two fingers toward each other until they meet in the middle. That is approximately the pattern of a doublet antenna

Antenna gain plots are almost always in dBi, so here is the doublet plot in dBi

Doublet Gain Plot in dBi

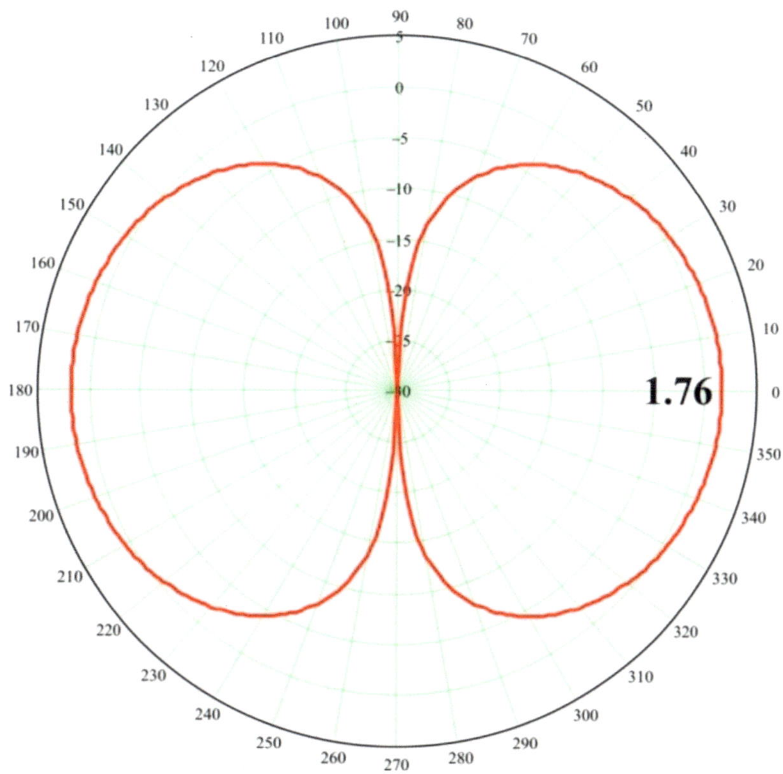

[17]Here is something to remember. Text books give antenna equations in voltage. But, the gain of an antenna and the plots displaying it are in power, which is normalized to voltage squared.

A Real Dipole

The dipole is the simplest antenna. It can be any length, but the most common are the one-half wavelength dipole and the very short dipole. The very short dipole is usually associated with the theoretical analysis, not a physical antenna.

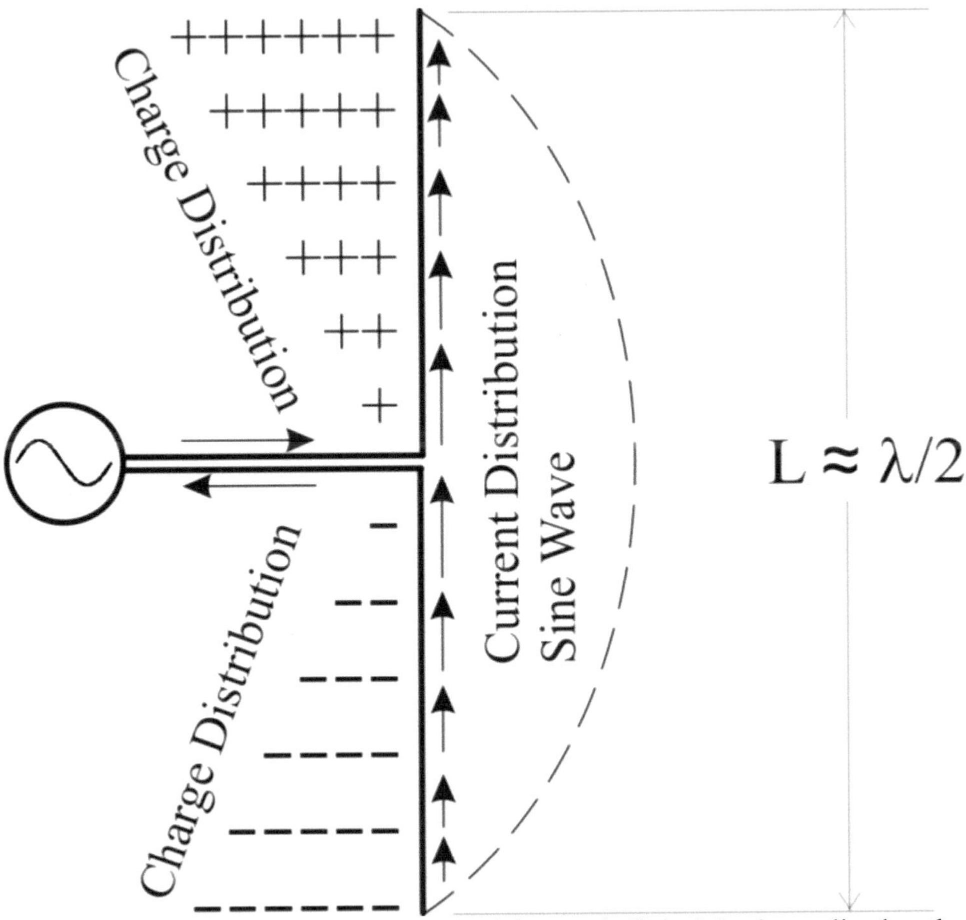

With a very (almost infinitely) short dipole, the current in the antenna is almost uniform. However, the half-wave dipole has a sine wave distribution over its length. As shown, there is also a distribution of charges.

The directive gain is 1.64 compared to 1.5 for the doublet. Expressed in dB, it is 2.15 dBi. Here are the plots of the two types. In practice you can use short dipoles, but they are very inefficient.[18] But there are applications where a short dipole can satisfy a requirement. There is no particular limit on how short it can be. The challenge is to match the receiver to the low impedance of the short dipole[19].

Dipole Gains in dBi

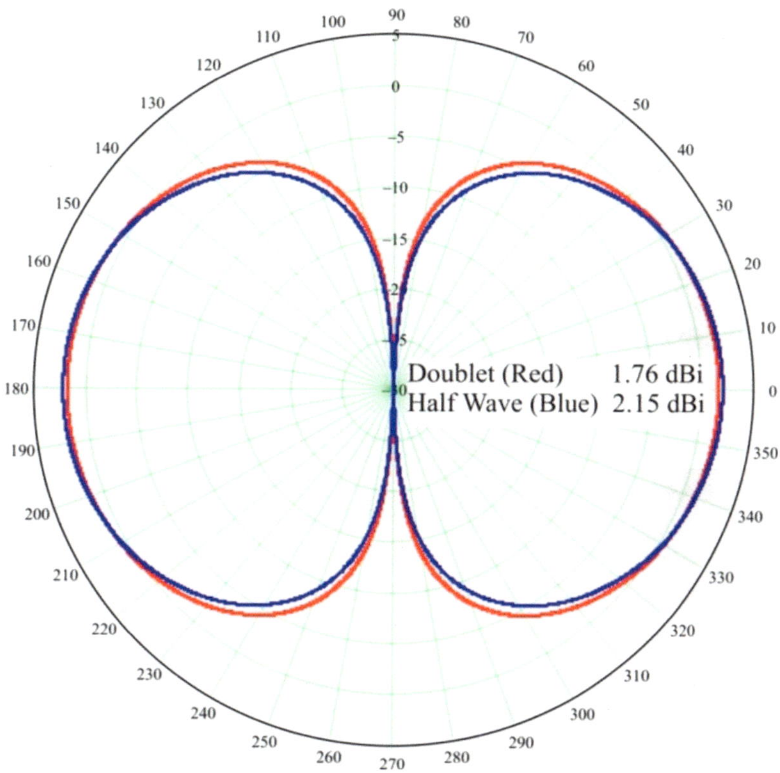

[18] Remember that "short" is relative to the wavelength of the driving signal.

[19] It is not common to use a transmitter with a short dipole because of the signal loss. Receivers can tolerate larger losses and still be useful.

Complex Impedance and Ohm's Law

If you are not knowledgeable of complex impedance, please go to the Appendix B for a tutorial.

Ohm's law applies to all three of the types of applications; i.e., direct current, alternating current and electromagnetic waves (EM). It is the fact that EM obeys Ohm's law that takes much of the mystery (black magic) out of it.

Non-Linearity, Signal Modulation and Intermodulation

I previously stated that EM methodology involves the summation of discrete sine waves. Once you have fully accepted this, understanding EM processes becomes much clearer and the mathematics of EM is simplified. An EM frequency is a sine wave of an exact, single frequency. It has a zero bandwidth. No matter how closely you place another frequency to it, it is a separate frequency and the two frequencies to not interact in any way. But, there are ways that single or multiple frequencies can give rise to **additional discrete** sine waves.

Linearity

Linearity means that a graph of the input versus the output is a straight line. An example is applying a voltage to a resistor and measuring the current. The plot is a straight line. Another

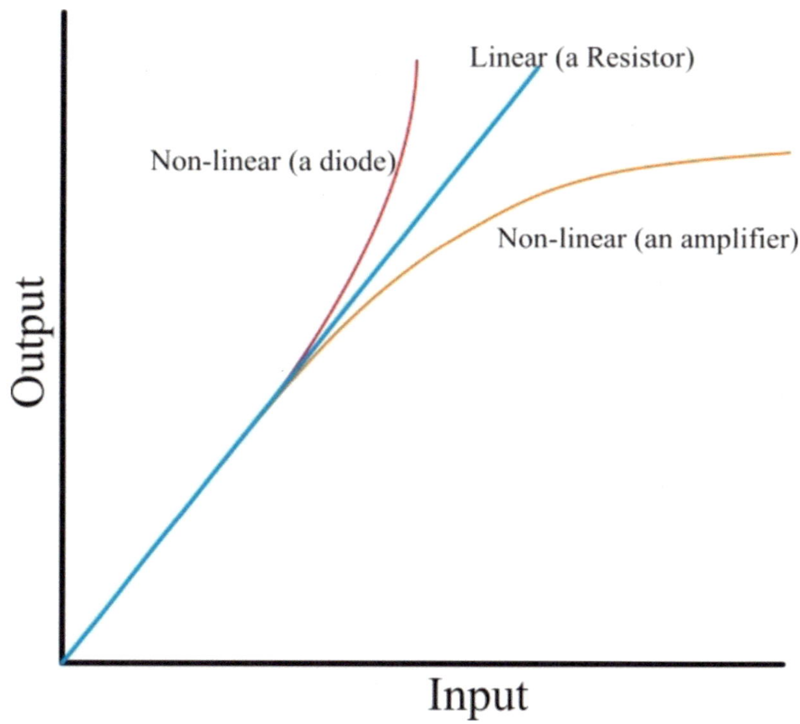

example is driving an amplifier at very low level. This is not as perfect as the resistor example, but it is "close enough". In a linear circuit the frequency you put in is the only frequency you get out.[20]
The figure above shows typical input/output responses.

Non-Linearity, Harmonics
Non-linearity generates harmonics. The desired signal is the "fundamental", fo, and that is the desired output. But, in addition, you get 2∗fo, 3∗fo, 4∗fo and so on. Usually, only the first 3 or 4 are troublesome. But note that the harmonics are each single frequency, zero bandwidth signals that are exact multiples of the fundamental.

Non-Linearity, Intermodulation Mixing
Now consider amplifying two or more CW tones simultaneously with a non-linear device. The f1 and f2 will both generate integer harmonics of their frequencies. But, in addition, all of the tones will mix together to yield other single frequencies. Examples: f1±f2, 2f1±3f2,etc.. So, the number of combinations is huge. But every one of them is a

---

[20] We are in the business of transferring information. A single frequency (called CW for "continuous wave" or "tone") that has been on for all time and will never be turned off contains no information. Information is introduced by **modulation**. Even turning the CW ON and OFF is a form of modulation; i.e., Morse code. Let's take a bizarre example: A spy ring needs to know when an office is empty. A CW transmission signal is turned on. Three weeks later the first spy observes that the office is finally empty and turns the transmission off. The second spy observes the absence of the CW signal and knows it is a go; voila, modulation has occurred.

single frequency, zero band width tone. And there are no frequencies between theses discrete tones. Then, if there are even more "fundamentals" the number intermodulation tones becomes enormous. This effect represents a major problem in the operating a transmitter amplifier in its non-linear region. The intermodulation tones cause distortion[21] and possibly issues with generating frequencies not authorized by the FCC.

---

[21] Both of these distortions also happen at audio frequencies. That is why "purists" pay big bucks for a linear audio amplifier to play music with excellent clarity.

Time and Frequency Domains

There two ways to represent individual sine waves. They are mathematically related to each other by the Fourier transform. See Appendix B for an example.

Time Domain
In the time domain plots, the x-axis is time and the y-axis is voltage (or current).

Frequency Domain
In the frequency domain, the x-axis is frequency and the y-axis is relative power in dB. It presents as zero width spikes on the plot. Notice that, for this case, fo = 0. The fundamental frequency carrying the modulation is not even present!!

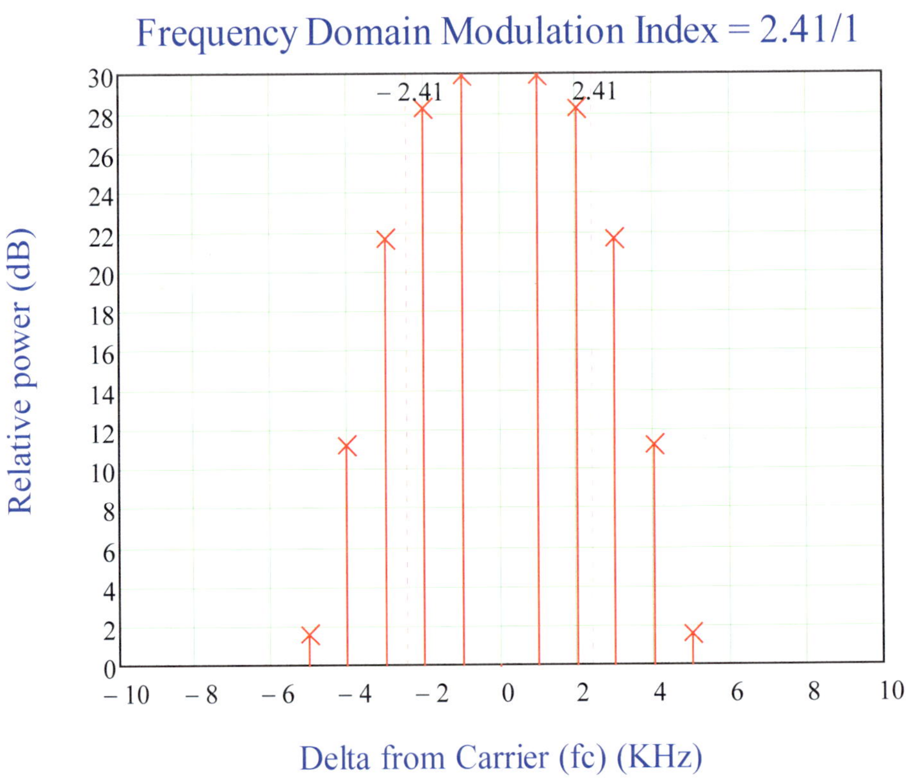

Reality Subjects

Periodic Signals

Are you ready for another miracle? Envision a segment of time called a period, usually designated as "T". Now, envision a voltage vs time signal of **any** shape that occupies that period. If the voltage signal repeats exactly over all periods, the signal can be decomposed into a series of discrete, individual sine waves. The lowest frequency sine wave is the "fundamental" and it is equal to 1/T. All of the other sine waves are harmonics, meaning integer multiples of the fundamental i.e., 1f, 2f, 3f, 4f, 5f… out to infinity. The name of this effect is "Fourier Series".

The task of finding the amplitude and phase of each harmonic is advanced mathematics and is beyond the scope of this book. Virtually any EM text book will show you how to do it. But, be warned that they may have different methods that yield the same answer.

I have chosen the very simple square wave as the vehicle to illustrate the Fourier series. The period, T, is 1.0 millisecond, so the fundamental frequency is 1 KHz. A characteristic of the square wave is that a summation operator can be used to find the harmonic values[22]. Also, the frequencies, for this case, contain only the odd harmonics; 1f, 3f, 5f, 7f, 9f…….

So, here are the equations for the harmonics of the square wave.

[22] This is not so for some irregular waveforms. The components must be determined individually.

$$T \equiv 1 \cdot ms \qquad f \equiv \frac{1}{T} = 1\,KHz$$

$$Sum5 \equiv \frac{4}{\pi} \cdot \sum_{k=1}^{5} \frac{\sin[2\pi \cdot (2k-1) \cdot f \cdot t]}{(2k-1)}$$

$$Series5 \equiv \frac{4}{\pi} \cdot \left(\sin(2\pi \cdot f \cdot t) + \frac{1}{3} \cdot \sin(6\pi \cdot f \cdot t) + \frac{1}{5} \cdot \sin(10\pi \cdot f \cdot t) + \frac{1}{7} \cdot \sin(14\pi \cdot f \cdot t) + \frac{1}{9} \cdot \sin(18\pi \cdot f \cdot t) \right)$$

$$Sum20 \equiv \frac{4}{\pi} \cdot \sum_{k=1}^{20} \frac{\sin[2\pi \cdot (2k-1) \cdot f \cdot t]}{(2k-1)} \qquad Sum2000 \equiv \frac{4}{\pi} \cdot \sum_{k=1}^{2000} \frac{\sin[2\pi \cdot (2k-1) \cdot f \cdot t]}{(2k-1)}$$

The exact solution requires that the summation be taken to infinity, but that would take more time than I have!!

Here is plot of these equations:

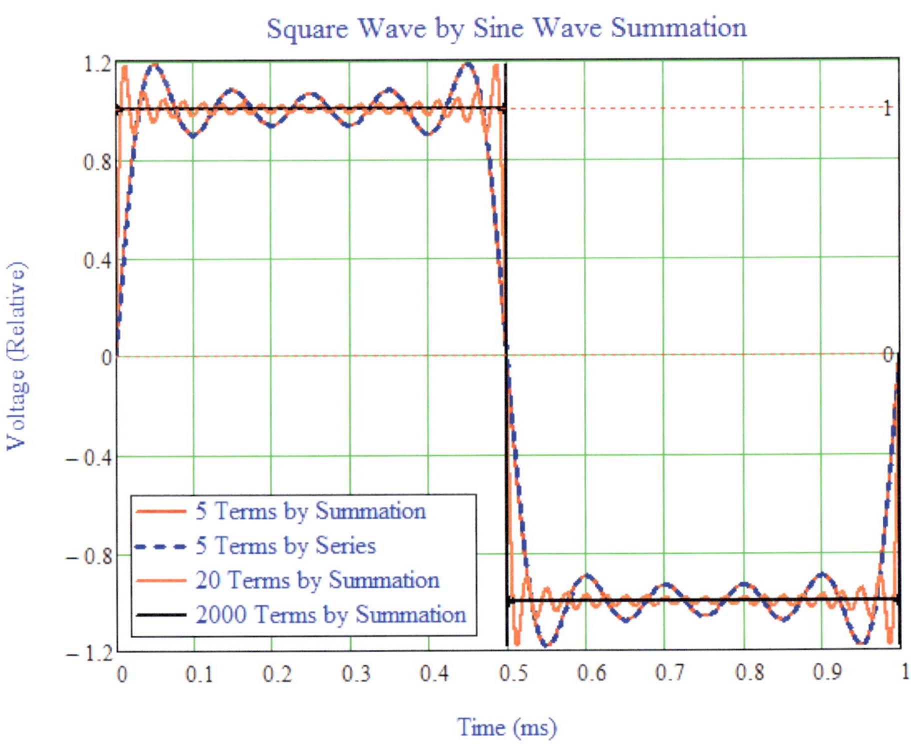

Note that the Sum5 and Series5 are identical. Also, observe that the approximation to the square wave gets better with higher number of harmonics.

The greatest error occurs at the transition from +1 to -1. Here is an expanded view of that area.

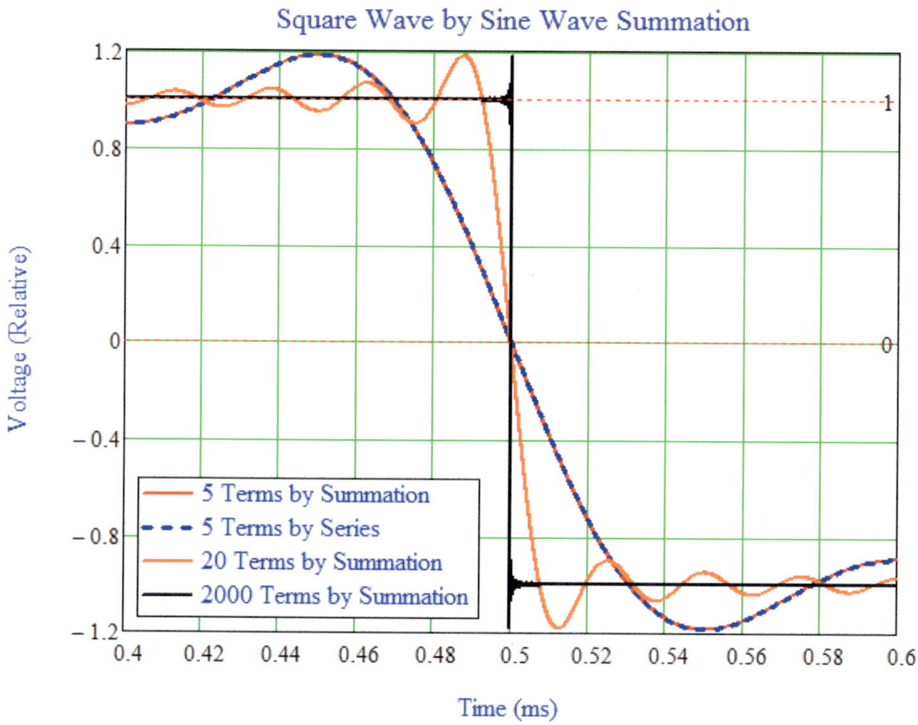

Side note:
I still remember the time, many decades ago, when I realized that periodic signals could be decomposed into the summation of individual, discrete sine waves. In my mind it bordered on miraculous. (Much in the same way that three colors can be combined to produce any other color.) I spent a long time in study trying to understand the equations. The situation was made worse because different authors approach getting the component values differently.

Here is the frequency domain plot of the first five sine waves.

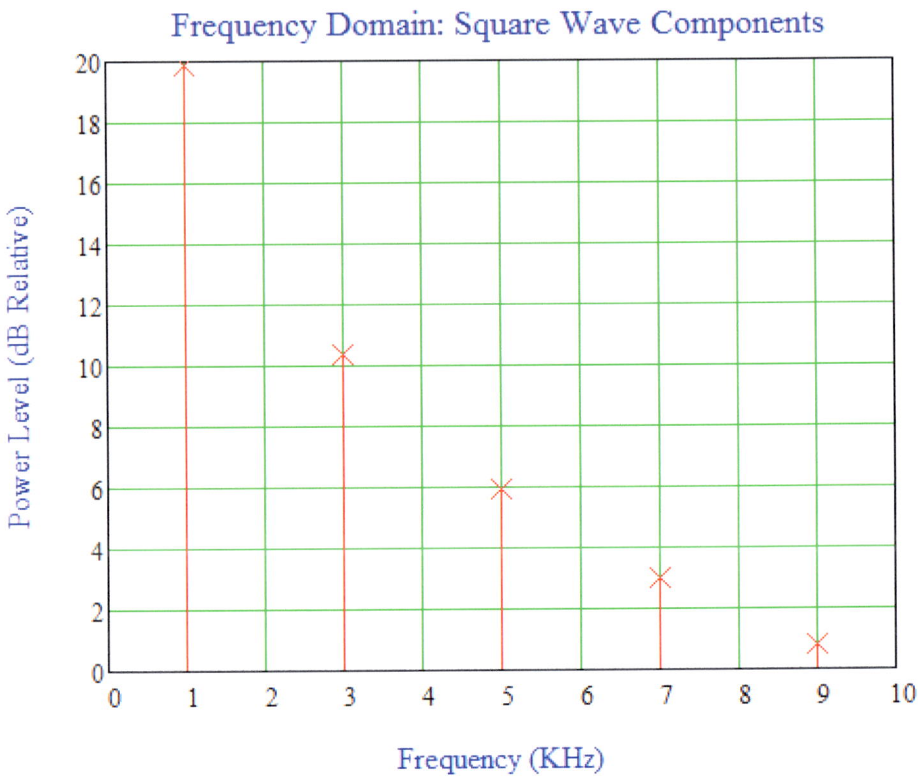

Once again this demonstrates the discrete nature of EM analysis. The summation of sine waves!!

Transmission Lines and Waveguides

Transmission lines move EM energy from one point to another point. There are many types of transmission lines, but they all have one thing in common. They confine the EM energy to the region near the transmission line and they transport the EM energy along the transmission line. The operational characteristics of the transmission line are very much dependent on its size relative to the wavelength of the signal. If the wavelength is very long (relatively), then the effect of the transmission line on the signal is minimal. At short wavelengths the transmission line is itself part of the circuit and its effect must be considered in the overall design.

There are two types of transmission lines:

Two conductor transmission line

There is one wire for the outgoing current and one for the returning current. These transmission lines operate in the TEM mode. The "T" is for transverse (perpendicular) to the direction of wave propagation. The "E" is electric field and the "M" is magnetic field. Mode TEM means that both fields are transverse to the direction of propagation. The speed of propagation is equal to the speed of light unless the structure includes a dielectric in its construction. The speed is the same for all frequencies. Examples: Twin lead transmission lines usually have a plastic web holding the conductors in place. Coaxial transmission lines usually have a dielectric material separating the inner conductor from the outer conductor. In both cases the presence of the dielectric material reduces the speed of propagation to be less than the speed of light. A typical coaxial cable slows the propagation to about 70% if the speed of light.

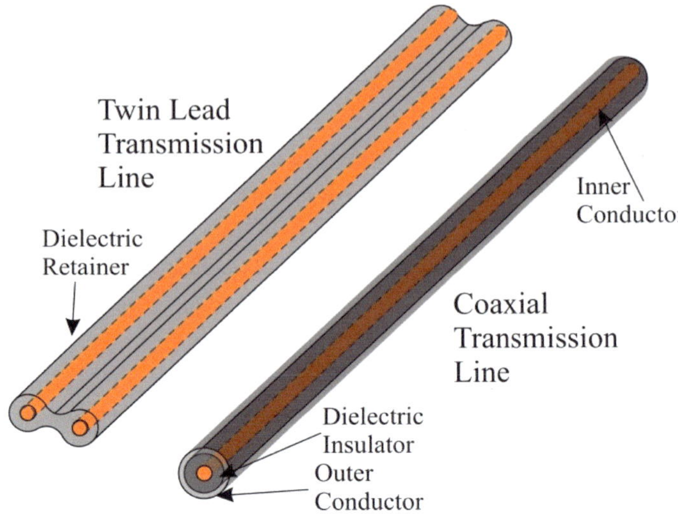

In the coaxial transmission line the EM wave is contained inside between the inner and outer conductors. There is no current on the outside of the outer conductor. Therefore, the cable can be near other cables or objects without interacting. On the other hand, the E and H field surround the space around the twin leads. So, nearby objects will interfere with the transmission line. If you construct a short piece of wire and placed it near and parallel to one of the wires, you can extract energy from the twin lead. It would also disturb the impedance of the twin lead at that point. The spacing between the twin leads must be much, much less than a wavelength. The current in the two wires is 180° apart, so in the far field, all of the voltages sum to near zero. Very little energy is lost.

Single conductor transmission

This type is more usually termed waveguide. It is a hollow tube of any cross section, but usually rectangular or circular. The EM energy is injected into the waveguide at one end and travels to the other end. It operates in either the TE mode or the TM mode[23]. The TE mode is more commonly used. The energy does not progress straight through. It transports by reflections off of the opposite walls. Different wavelengths bounce off the walls at different angles. Therefore, the speed of propagation depends on the frequency.

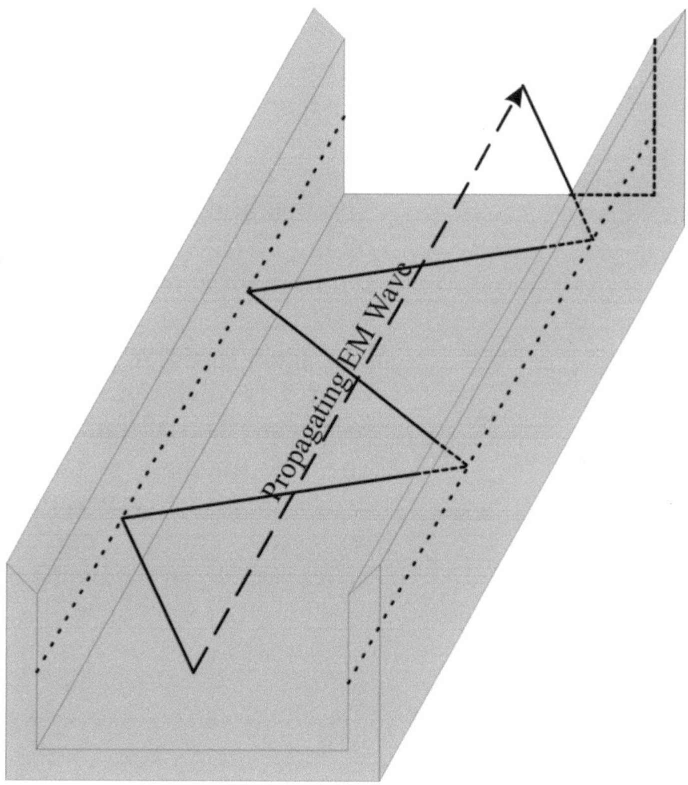

[23] In each case the opposite, TE or TM, is parallel to the direction of propagation.

Pulsing energy, such as a square wave, is actually composed of many harmonically related sine waves, all traveling at different speeds. So, when the nice input square pulse comes out the other end, it is no longer square, it has rounded shoulders. This effect is called "dispersion" and it greatly limits waveguide operation. The longer the waveguide run, the worse the dispersion.[24]

The great thing about waveguide is that the only loss is the resistance of the walls. If you plate the inside with silver, the loss can be very low. Aluminum is a common material for waveguide; but, compared to silver, it is very lossy, but also much less expensive. Some wave guides are constructed from copper, which is almost as good as silver.

Cutoff Frequency

The operating range for waveguide is determined by the distance from side-to-side. If that distance is one half a wavelength, the wave is at **cutoff**. Wavelengths longer, frequencies less, will not propagate through the waveguide. If the distance is greater than one wavelength, the wave will propagate through but may break into multiple TE or TM modes. This causes unpredictable distortion of the waveform at the output. A single waveguide normally operates over slightly less an octave range, an octave being 2-to-1.

---

[24] Fiber optic cable is also a form of waveguide except the medium is glass instead of air. Just think how purely transparent the glass must be to pass the signal so far. Even so, an optical amplifier must be inserted every so often to maintain signal strength.

Below Cutoff

I am betting that in school you were not taught what happens below the cutoff frequency. Let's look first at a frequency that is just barely above cutoff. It will propagate, but very slowly, because the angles of the reflections are almost perpendicular to the walls. At frequencies below cutoff, the input to the waveguide appears as a reactive component. For TE waves the waveguide emulates an inductor. For TM waves the waveguide emulates a capacitor. Both operate by accepting energy in the first half period and returning the energy during the second half period. The closer the frequency is to cutoff, the further it goes into the waveguide before coming back out.

Let's consider how an inductor works. Energy goes out, then it comes back; no power is transmitted. But, bring another inductor (coil) within the region of the first inductor and second inductor receives power. This is a transformer!

Now, apply the inductor analogy to the waveguide below cutoff. Many authors state simply that no power is transmitted below the cutoff frequency. But there is power inside the waveguide near the mouth. The wave is evanescent, meaning attenuating (decaying) and in a long piece of waveguide it will die out. But with a short piece of waveguide you can recover some of the power in the analogous manner of the second inductor. The depth of penetration of the energy is a function of how far removed the operating frequency is from the cutoff frequency.

A very good graphical presentation of the property of the evanescent wave in waveguide is displayed on a website by Dr. Daniel A. Russell. Surf the web for “Evanescent Waves in Waveguides” and select “Waveguides and Evanescent Modes”. Dr. Russell’s animation is for acoustic waves, but the general effect is the same for electromagnetic waves; except that the n=0 mode is not possible for the EM waves in metal waveguide. On the animation set the mode to 1, and then use the slider to set the normalized frequency above and below 1.

Here are the equations for how much the evanescent signal attenuates as it travels down the waveguide.

$$\lambda_{oc} \equiv \frac{c}{1 \cdot GHz} = 0.3\,m$$

$$\lambda_o \equiv \frac{c}{800MHz} = 0.375\,m$$

$$\alpha \equiv \frac{2\pi}{\lambda_o} \cdot \sqrt{\left(\frac{\lambda_o}{\lambda_{oc}}\right)^2 - 1} = 12.6\,\frac{1}{m} \quad \text{Nepers/meter}$$

$$\alpha_{dB} \equiv \frac{2\pi \cdot 8.69}{\lambda_o} \cdot \sqrt{\left(\frac{\lambda_o}{\lambda_{oc}}\right)^2 - 1} = 109.3\,\frac{1}{m}$$

Multiply by 8.69 to get dB/m

λ_{oc} is the cutoff wavelength

λ_o is the operating wavelength

In this example:
Cutoff frequency is 1 GHz
Operating frequency is 800 MHz
The attenuation is 109.3 dB/m = 2.8 dB/inch

So, if you make a waveguide section two inches long, it will attenuate the 800 MHz signal by 5.6

dB[25]. And here is the kicker, there is no phase shift!!

Characteristic Impedance

All forms of EM transport, including free space, have an associated characteristic impedance. (I usually just say impedance.) The usual symbols for impedance are: characteristic impedance Zo, general impedance Z, impedance at the input of the transmission line Zi and the impedance of the termination of the transmission line Z_L

Transmission line equations

As I stated previously, when an object becomes larger than one wavelength, things change. This is especially true for transmission lines. Consider this transmission line L wavelengths long.[26].

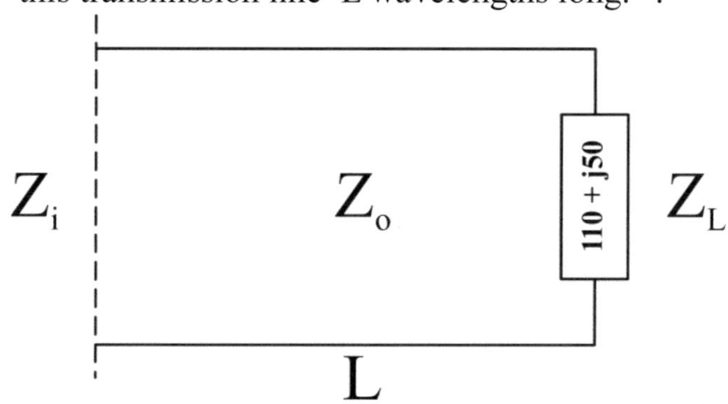

The horizontal parallel lines represent any transmission line; twin lead, coaxial, etc. Z_L is the impedance of the load. Z_o is the characteristic impedance of the transmission line. L is the length of line from the load to the point of

---

[25] So, it is not strictly correct to say that a waveguide will not propagate a signal below cutoff. It will, but with considerable attenuation.

[26] This is wavelengths in the transmission line. Remember that the presence of a dielectric causes wavelengths shorter than in free space.

observation. Z_i is the impedance at the observation point. The input impedance, Z_i,, changes as the length (L) changes. Here is the equation for Z_i;

$$Z_i(L) \equiv Z_o \cdot \left(\frac{Z_L + j \cdot Z_o \cdot \tan(2\pi \cdot L)}{Z_o + j \cdot Z_L \cdot \tan(2\pi \cdot L)} \right)$$

where L is expressed in wavelengths. Sometimes you will see it like this: $\beta = 2\pi/\lambda$ and $\tan(\beta * L)$ where L has the same dimensions as λ This equation is for a lossless line and that is adequate for most situations.[27]

Observe these results for the case shown above:

$Z_o = 50$

$Z_L = 150 + 50j$

$Z_i(0.25) = 15 - 5j$

$Z_i(0.50) = 150 + 50j$

$Z_i(0.75) = 15 - 5j$

$Z_i(1.00) = 150 + 50j$

Notice two things:

The values repeat every one half wavelength

At one half wavelengths the input impedance is the same as the terminating impedance. This is true for any value of the transmission line characteristic impedance!!

When L is ¼ wavelengths, an open looks like short and vise-versa

---

[27] See Appendix C for the case of lossy transmission line.

Quarter Wave Transformer

Here is a neat little property of transmission lines. Suppose you have a 50 ohms source driving a 50 ohms transmission line. But you would like connect to two 50 ohms lines in parallel, making the impedance 25 ohms. That gives a VSWR of 2.0 to 1. You can insert section of transmission line between the two and match back to VSWR of 1.0 to 1.

The equation for the transforming transmission line impedance is the geometric mean of the impedance on the input side of the transformer and the impedance on the output side of the transformer. In our case, 35.355 ohms.;

$$Z_t \equiv \sqrt{50 \cdot 25} = 35.355$$

$$Z_o = 35.355$$

$$Z_i(0.25) = 50$$

Observe that Z_i checks out to be 50 ohms!!

Please remember that all of this is relative to wavelength, a function of frequency, which means that the solutions are only true at a single frequency.

Another Equation

The square root of the product of impedance of the transmission line terminated with a short ($Z_L = 0$) and the impedance of the transmission line terminated with an open ($Z_L = \infty$) equals the characteristic impedance of the transmission line. Except when $L = 0$, the product of $0 * \infty$ is undeterminable, and the impedance is Z_L.

$Z_L = \text{Infinity}$

$$Z_o \cdot \left(\frac{1}{j \cdot \tan(2\pi \cdot L)} \right)$$

$Z_L = \text{Zero}$

$$Z_o \cdot (j \cdot \tan(2\pi \cdot L))$$

$$\sqrt{Z_o \cdot (j \cdot \tan(2\pi \cdot L)) \cdot \left[Z_o \cdot \left(\frac{1}{j \cdot \tan(2\pi \cdot L)} \right) \right]}$$

$$\sqrt{Z_o^{\,2}} = Z_o$$

But, before you get too excited about this formula let me tell you that I have tried to use it, but with poor results. The problem is that you cannot get a true open. If you cut the end of a transmission line, the EM wave does not stop instantly at the end. There are fringing fields off the end of the transmission line and maybe a tiny bit of actual transmission. In my case, it was enough to make the calculation useless.

Standing Waves

This is concept that I had a hard time getting my mind around. The key is in remembering my previous statements that EM sine waves of the **same frequency** do not mix in space, but they do **sum together**, even if they are **not going in the same direction.** In fact, a standing wave is produced when a wave and its reflection occupy the same space at the same time.

Consider two space sine waves: One traveling toward the barrier and a reflected one traveling away from the barrier. Since both occupy the same space at the same time, their voltages add together. This creates a "standing wave" that is **stationary in space**.

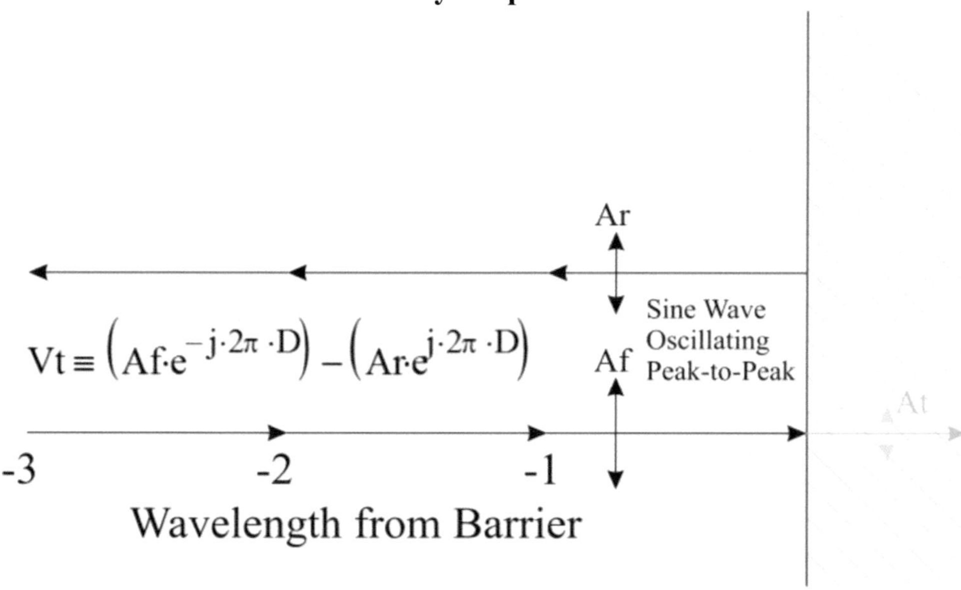

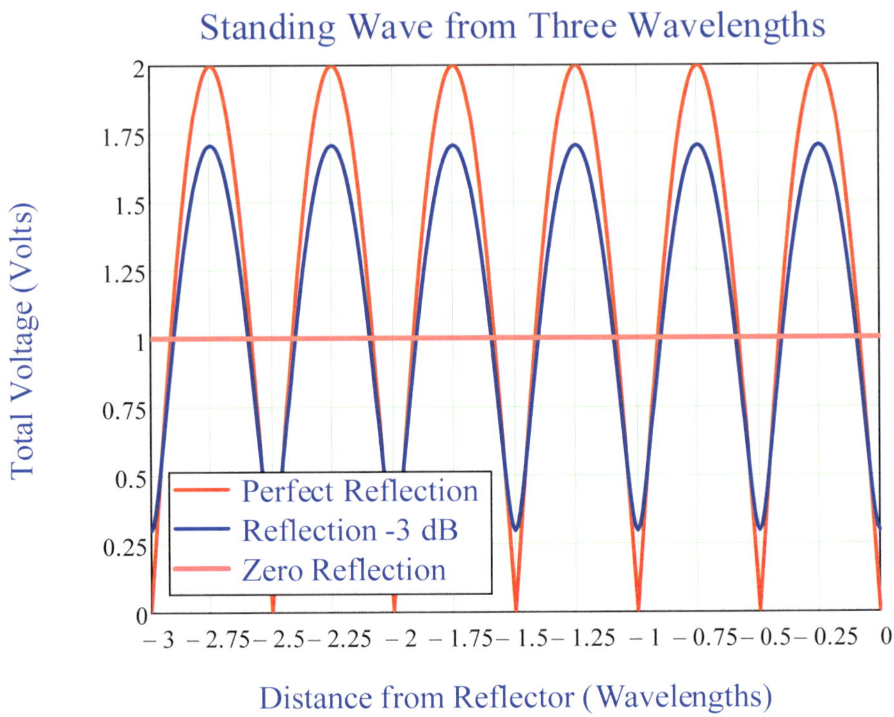

Actually, the term "standing wave" is a misnomer. The plot shows the result of summing the two space waves together. Observe the equation. The minus sign between the terms comes from a 180^0 phase shift at the barrier. When the forward going wave hits the barrier, it cannot just vanish. It must either bounce off of it (reflect) or enter it (transmit). Actually, to a greater or lesser degree, both things happen. For a nearly perfect conductor such as copper, virtually all of the energy is reflected, as indicated by the red plot. The summation creates a condition in space where the summed sine waves create **stationary peaks and valleys**. If you place a matched antenna at -0.25λ away from the barrier, you get maximum received power. If you place it at -0.5λ, you get minimum or zero power. Inside the barrier the transmitted sine wave travels along at constant level, no standing wave.[28] The fact that the exponent of e in the second term is positive indicates it is traveling in the opposite direction. This same thing happens in waveguides.[29]

Sometimes a once in a lifetime event occurs. For me, it was on the beach at Corpus Christy, Texas. I observed a shallow wave coming inward from the gulf and moving up gently sloping sand. At the same time there was water that had gone all the way up and was coming back. At the point of observation was a "wrinkle" in the water; a place where the water formed a **stationary crest** above the sand that was shaped like half a sine wave. I shouted to my family, "Come look, come look, here is an actual standing wave." If I lived near the ocean I probably could have observed this many times. But I don't and it was a real thrill to see it.[30]

---

[28] This is true only if the transmitted wave has dissipated all of its power before it exits the barrier. If not, another reflection occurs at the exit point and the whole situation changes drastically. I will cover this in the section on radomes.

[29] Back when I was a pup, we used slotted waveguide to measure the power level. The distance from peak to valley was 0.25λ. Of course that was wavelength in the waveguide, not free space. Thank Hewlett Packard for the EM engineer's best friend, the network analyzer!

[30] I realize that this is not quite the same thing because the laminar flow water is not a wave, but is was close enough for me!

VSWR

The ratio Vmax/Vmin is called the Voltage Standing Wave Ratio (VSWR) expressed as xx:1. A VSWR of 1.0:1 means no reflection. A VSWR of ∞:1 means total reflection. Antennas and other devices are characterized by their VSWR.

On a transmission line when the forward going wave, vf, encounters the antenna, some of it is reflected and travels back in the reverse direction, vr. The ratio vr/vf is the reflection coefficient and it is a complex number with magnitude ≤ 1.0.

$$\frac{vr}{vf} \equiv \frac{Z_L - Z_0}{Z_L + Z_0} = \rho$$

Where vr and vf are voltages, ρ is the reflection coefficient, Z_L is the impedance of the antenna and Z_0 is the characteristic impedance of the transmission line. All of the parameters may be complex (a + jb), but Z_0 generally is not complex.

The VSWR is:

$$VSWR \equiv \frac{1 + |\rho|}{1 - |\rho|}$$

The net effect of the reflected wave is a loss of power to the antenna.

When the EM signal reaches the antenna three, and only three, things happen. Some is reflected, some is consumed as heat in the antenna and the remainder is radiated. But it may not be radiating in the direction and manner you want. Just because an antenna has “good VSWR” does not mean it is a good antenna. This is especially important in the case where a manufacturer specifies an upper frequency limit and the VSWR looks good above that frequency

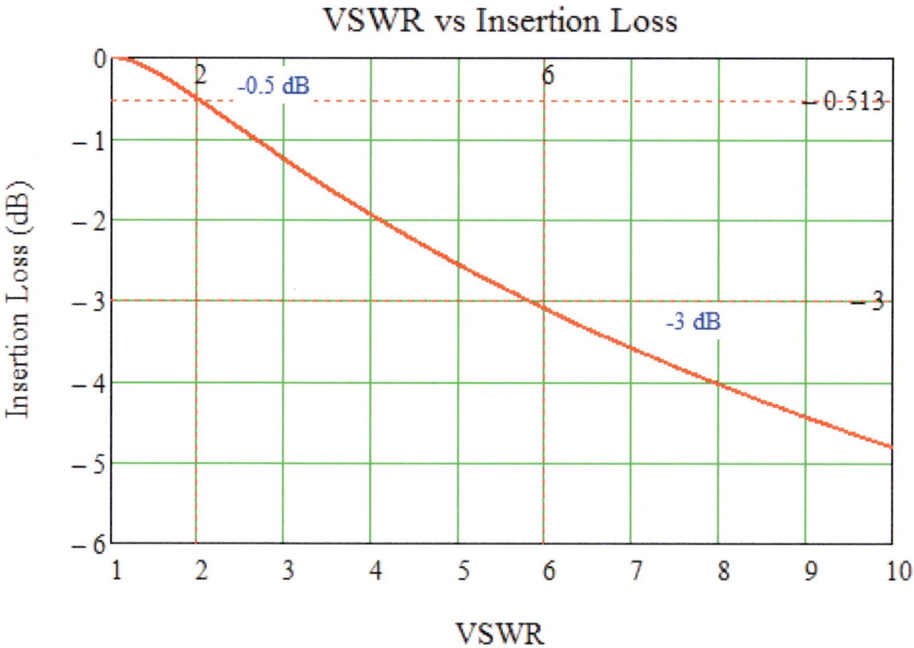

There are two things about this plot that I want to emphasis.

First, sometimes antenna purchase specifications are written by persons with minimal experience with antennas. They specify the maximum VSWR as 2.0:1. That represents a loss of only 0.5 dB. It is really difficult to design antennas with less that 2.0:1. Does the customer really need it to be that good. Sometimes, yes, but more likely, no. I am generally satisfied with less than 3.0:1 (1.2 dB loss).

Second, I was surprised when I realized that a loss of half the power (3 dB) occurs at approximately VSWR = 6:1

Noise Figure and G/T

Here is another miracle. We think of a resistor as passive, but, in fact, it is a noise generator. At any temperature above absolute zero degrees Kelvin (-273.15 degrees Celsius), it has a voltage at its terminals.

Equivalent circuit of resistor with "noise generator"

At matched the condition i.e., maximum power transfer, it has the form:

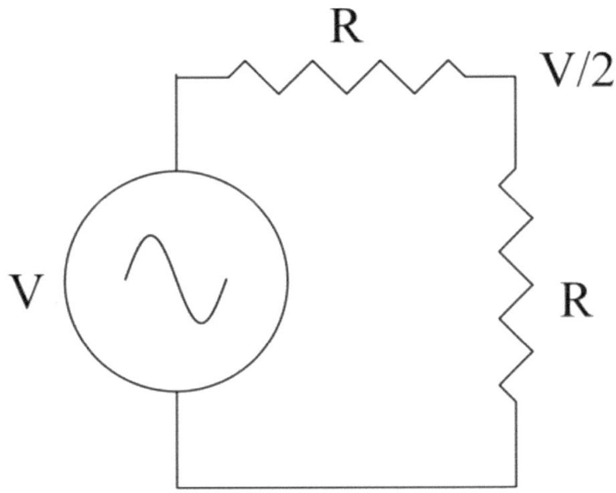

The RMS voltage generated by the resistor is:

Resistor noise voltage $=\sqrt{4kTBR}$
where
k = Boltzman's constant = $1.38*10^{-23}$ Joules/K
T = Temperature in K
B = Bandwidth in Hz
R = Resistance in Ohms

Attach the resistor to a matched power meter.

Power Meter

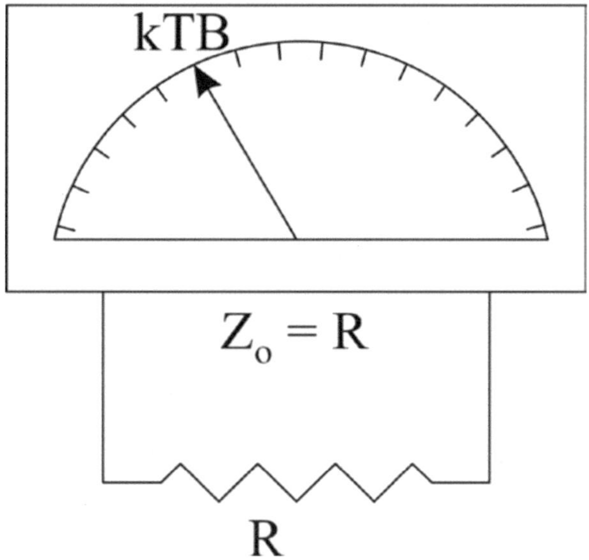

And, you get:

Resistor RMS noise voltage = 4kTBR
where
k = Boltzman's constant = $1.38*10^{-23}$ Joules/K
T = Temperature in K
B = Bandwidth in Hz
R = Resistance in Ohms

The B is the bandwidth of the power meter. The value displayed on the meter is KTB[31]. The bandwidth of a typical spectrum analyzer can be set from about 1 Hz to 3 MHz, or more.

The noise levels for three bandwidths relative to one milliwatt are shown here:

---

[31] You often hear that the noise value of a resistor is KTB. But that is true only in the case of a matched circuit as shown here.

$$10 \cdot \log\left(\frac{k \cdot 293.15 \cdot K \cdot 1 \cdot Hz}{1 \cdot mW}\right) = -173.9 \quad \text{dBm}$$

$$10 \cdot \log\left(\frac{k \cdot 293.15 \cdot K \cdot 10 \cdot KHz}{1 \cdot mW}\right) = -133.9 \quad \text{dBm}$$

$$10 \cdot \log\left(\frac{k \cdot 293.15 \cdot K \cdot 1 \cdot MHz}{1 \cdot mW}\right) = -113.9 \quad \text{dBm}$$

For 290K (62.3 degrees F) and a bandwidth of 1 Hz the noise voltage into a 50 ohms load is 0.0009μV.

A receive/transmit system adds noise to the desired signal. This noise sets the lower limit of the signal that can be transmitted and recovered.

Every receiving system has three components in common:

1- An antenna
2- A loss from the antenna to the receiver; i.e., cable loss
3- At least one stage to receive the RF energy

All of these contribute some additional noise into the system. However, in most cases the loss in the antenna is very small and can be ignored.

The third component attracts the most attention. It is usually an amplifier and its figure of merit is Noise Factor (F). You probably hear the term Noise Figure (NF) more often. The noise figure is just the noise factor express in dB; i.e., NF = 10 log(F)

The signal coming into the amplifier has noise associated with it (Ni). (More on that later.) It is also receiving a desired signal (Si). So the ratio Si/Ni is the input signal-to-noise ratio. But, every receiving system adds its own noise, so at the output the signal-to-noise ratio is So/No that is less than Si/Ni. The noise factor is the ratio of these two ratios: F = (Si/Ni)/(So/No).

And, to make things even more interesting, there is another way to define the noise contribution. It is noise temperature (Te) and it is the physical temperature the component would have to be to generate the same level of noise as the noise factor. They are two peas in the same pod, except they are different color.

The equation relating noise factor to noise temperature is: Te = (F-1)*To, where To is the actual thermal temperature of the component; usually taken as 290K, assuming room temperature operation.

If the physical temperature of the device is not the same as the room temperature, its noise temperature changes.

Noise temperature of a component with physical temperature of T_{Phys}. is:

$$T_e = (L - 1)\cdot T_{Phys}$$

Where L is attenuation ratio.
To convert from dB use: $L(NF) \equiv 10^{\frac{NF}{10}}$

Then the noise factor is:

$$F(NF, T_{Phys}) \equiv 1 + \frac{(L(NF) - 1)\cdot T_{Phys}}{T_{Ref}}$$

$F(2,290) = 1.585$

$F(2,50) = 1.101$

and the Noise Figure is:

$$F_{dB}(L, T_{Phys}) \equiv 10\cdot \log(F(L, T_{Phys}))$$

$F_{dB}(2,290) = 2$

$F_{dB}(2,50) = 0.417$

In this example the noise figure at room temperature is 2 dB and the noise factor is 1.585.

Cool the device down to 50K and the noise figure drops to 0.417 and the noise factor drops to 1.101. This is why astronomers cool the waveguide and the first amplifier as much as they can. This same effect is true for receiver noise figure.

Cascaded Operation

In a receiver with multiple stages, each stage has its own noise factor. There are two equations for a cascade. One is for noise factor and the other is noise temperature. The advantage of the noise temperature equation is that the components can be at different physical temperatures.

Here is an example of a five stage receiver. All of the components are at reference temperature of 290K.

Convert Noise Figure to Noise Factor

$$NF(Fig) \equiv 10^{\frac{Fig}{10}}$$

Convert Gain in dB to Gain in Ratio

$$G(dB) \equiv 10^{\frac{dB}{10}}$$

Convert Noise Figure in dB to Temperature degrees K

$$NT(NFdB) \equiv T_{Ref} \cdot \left(10^{\frac{NFdB}{10}} - 1 \right)$$

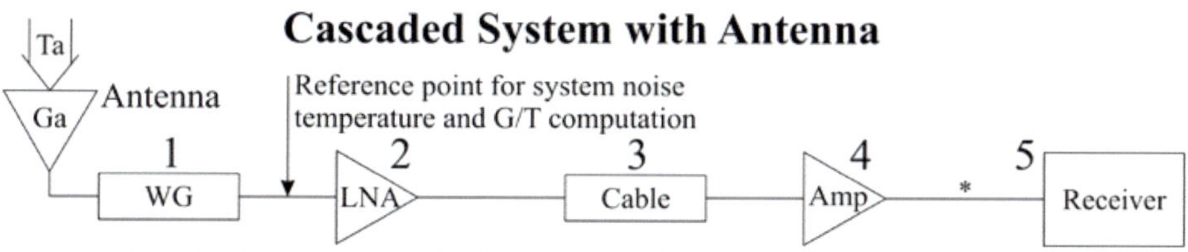

1	2	3	4	5
Gain = -1 dB	Gain = 20 dB	Gain = -5 dB	Gain = 15 dB	Gain = 0 dB
Gain = 0.7943	Gain = 100	Gain = 0.316	Gain = 31.6	Gain = 1
	NFig = 3 dB	NFig = 5 dB	NFig = 8 dB	NFig = 10 dB
	NFac = 2	NFac = 3.16	NFac = 6.31	NFac = 31.62
	NTemp = 288.6K	NTemp = 627K	NTemp = 1540K	NTemp = 2610K

Loss 1 includes waveguide or cable loss connecting the antenna to the LNA. If there is any heat dissipating loss in the antenna, it can also be included here.
*For the purpose of this example, the connection is assumed lossless.

Note that I have included both the ratio values and the dB values for each stage. Also, note that Loss is the inverse of Gain. So, the loss for stage 1 is 1/0.79 = 1.26; i.e., the loss equals the noise factor. **At room temperature the noise figure (or noise factor) of an attenuator is equal to its loss.**

The reference point for G/T computations is between the waveguide (or cable) and the input to the LNA

First solve for the noise figure and noise temperature of the LNA chain.

Compute Noise Factor and Noise Temperature of LNA chain

$$\text{NoiseFactor} \equiv \text{NF}(\text{F2}) + \frac{\text{NF}(\text{F3}) - 1}{\text{G}(\text{G2})} + \frac{\text{NF}(\text{F4}) - 1}{\text{G}(\text{G2})\cdot\text{G}(\text{G3})} + \frac{\text{NF}(\text{F5}) - 1}{\text{G}(\text{G2})\cdot\text{G}(\text{G3})\cdot\text{G}(\text{G4})}$$

$$\text{NoiseFactor} = 2.194$$

$$\text{NFg} \equiv 10\cdot\log(\text{NoiseFactor}) = 3.412 \qquad \text{NT}(\text{NFg}) = 346.2\ \text{K}$$

$$\text{NoiseTemperature} \equiv \text{NT}(3) + \frac{\text{NT}(5)}{\text{G}(\text{G2})} + \frac{\text{NT}(8)}{\text{G}(\text{G2})\cdot\text{G}(\text{G3})} + \frac{\text{NT}(10)}{\text{G}(\text{G2})\cdot\text{G}(\text{G3})\cdot\text{G}(\text{G4})}$$

$$\text{NoiseTemperature} = 346.2\ \text{K}$$

Observe that both methods yield the same answer, 346.2K.

G/T of a System

The G/T of a system also depends on the antenna gain, the background noise and the waveguide loss. So, let's get to those contributions now. The antenna temperature, Ta, is the noise radiated from objects within the beamwidth of antenna. Earth radiation is, by convention, assumed to be 290K. Deep space is about 5K in the frequency range of satellite communications. But, all antennas have some portion of the side lobes intersecting earth objects. So, a convenient temperature for this demonstration is 50K

Let our first example have antenna gain of 40 dBi and a noise temperature of 290K.

Compute G/T

Antenna noise temperature

$Ga \equiv 10000$ 40 dBi $Ta \equiv 290$

Noise from antenna	Noise from waveguide	Noise from LNA chain
$Ta \cdot (G(G1)) = 230.355$	$290 \cdot (1 - G(G1)) = 59.645$	$NT(NFg) = 346.199$

Total noise

$TN(Ta) \equiv Ta \cdot (G(G1)) + 290 \cdot (1 - G(G1)) + NT(NFg)$ $TN(Ta) = 636.199$

G/T is Gain at waveguide output/Total noise

$$GoverT \equiv 10 \cdot \log\left(\frac{Ga \cdot G(G1)}{TN(Ta)}\right) = 10.964 \quad \text{dB/K}$$

The noise of the LNA chain is still 346.2K. Now add the noise from the antenna and noise contribution of the waveguide.[32] The total noise is raised to 636.2K. The G/T is the gain at the output of the waveguide divided by the noise of the LNA chain. In this case G/T = 10.96 dB/K.

You all know that a matched antenna has an input impedance. **The input impedance of an antenna does not generate noise**.

You probably recognize the phrase **"attenuation added to the input raises the noise figure by the same amount"**. Let's test it by changing G1 from -1 dB to -2 dB.

$$GoverT \equiv 10 \cdot \log\left(\frac{Ga \cdot G(-2)}{TN(Ta)}\right) = 9.964 \quad \text{dB/K}$$

---

[32] Most antennas are constructed of metal, copper, bronze or aluminum. These have very low loss that can usually be ignored. Some antennas, such as a patch, contain dielectric or matching elements that have losses that cannot be ignored. This loss can be lumped into the waveguide loss.

The G/T decreased from 10.964 dB to 9.964dB, precisely a 1 dB change. It is important to note here that this is true only if the all the components of the cascade are at the same temperature, 290K. We are about to discover that marvelous things can happen if we can reduce the background noise and improve the LNA noise figure.

The 40 dBi antenna has a very small beamwidth. Point the antenna into free space, preferably straight up. In the frequency region for satellite communication, 1 GHz to 60 GHz, the "noise of free space" is about 5K, a lot lower than 290K. But in a practical sense all antennas have side lobes and getting completely free of earthly objects is very difficult. A more likely value for real satellite systems is in the 35K to 50K range

Use a 50K antenna temperature and the G/T improves to 12.5 dB/K. Now reduce the noise figure of the LNA to 1.2 dB. We see another increase to 15.0 dB/K. Now, reduce the waveguide loss to 0.25 dB. The G/T improves to 16.5 dB/K, an improvement of 1.5 dB, even though the leading loss was changed by only 0.75 dB. Now the rule about the leading loss changing by X dB causing the output S/N to change by X dB does not apply. In fact, reducing the noise ahead of the LNA can improve the output S/N by a dramatically higher value. Consider the huge improvement of output S/N, (G/T), if you also cool the waveguide and LNA, as the astronomer does. (There is a reason the deep space probe antennas are located in the desert and it is not for the comfort of the staff!!)[33]

---

[33] Surf the internet "G/T calculator" for some good on-line calculators.

Modulated Signals

The whole reason for concerning ourselves with EM is to transfer information from one place to another. In order to do that the transmitted frequency must be **modulated** by the information signal. There are many ways to "modulate" a carrier frequency. But, I will cover only the two most common types.

The first is amplitude modulation (AM) and the second is frequency modulation (FM). It is left as an exercise for you to research the many other forms of information modulation.

Though it is not much fun to sit around and listen to a one 1000 Hz, 1 KHz, tone, it is a good frequency to use to demonstrate the effect of modulation on the carrier.

First, let us consider AM. An AM signal is generated when a modulating circuit adjusts the level of the carrier frequency up and down in sync with the modulation frequency. These plots show a 10 KHz carrier frequency (fc) modulated by 1 KHz frequency (fm)[34]. The first plot shows the signal in "frequency domain". The second plots shows the "time domain:. In advanced mathematics, each one is the Fourier transform of the other.

Here is the equation of AM. I did not include a phase shift of fm for simplicity.

$$f \equiv \overrightarrow{[(1 + 0.5\cdot\sin(2\pi\cdot fm\cdot t\cdot deg))\cdot\sin(2\pi\cdot fc\cdot t\cdot deg)]}$$

[34] The carrier frequency is usually much higher that 10 KHz. I made it this low so you could better see the effect. The frequency domain spacing remains unchanged.

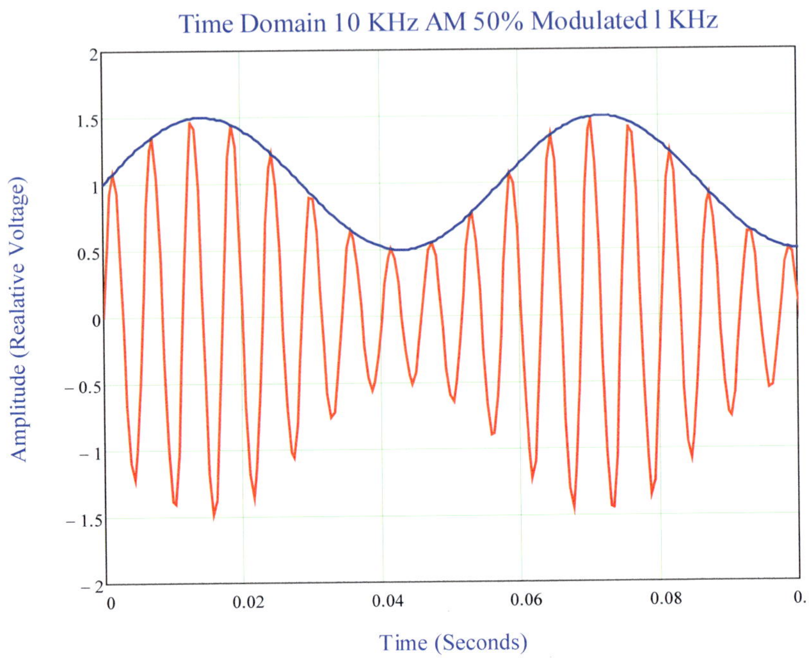
Time Domain 10 KHz AM 50% Modulated l KHz
Amplitude (Realative Voltage)
2
1.5
1
0.5
0
− 0.5
− 1
− 1.5
− 2
0
0.02
0.04
0.06
0.08
0.1
Time (Seconds)

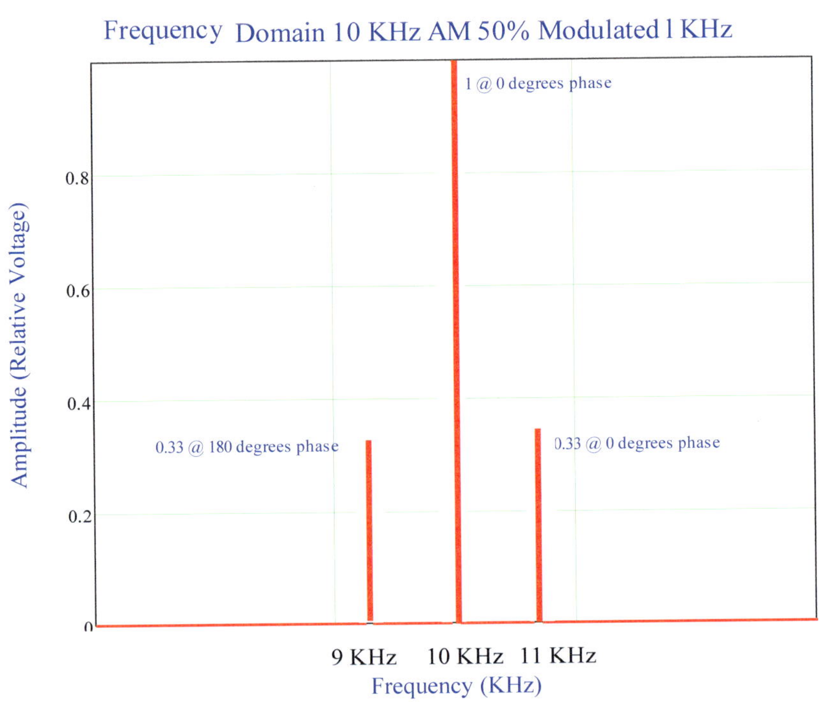
Frequency Domain 10 KHz AM 50% Modulated l KHz
Amplitude (Relative Voltage)
0.8
0.6
0.4
0.2
0
1 @ 0 degrees phase
0.33 @ 180 degrees phase
0.33 @ 0 degrees phase
9 KHz
10 KHz
11 KHz
Frequency (KHz)

Notice that there are three, and only three, discrete frequencies. Use three synchronized signal generators to sum the three frequencies in amplitude and phase to get the waveform shown in the time domain plot.

The second form is frequency modulation. In this case a circuit changes the frequency of the carrier, fc, in sync with the modulating tone. Let's again use 1 KHz for the modulating tone, fm. I will use the frequency domain plots only because the fm is so small compared to the fc that you really cannot see the changes. But, there is another variable, fd, that defines how much you are going to change the frequency. And from that we establish the ratio fd/fm, the modulation index. Depending on the value of the modulation index the frequency signal can classify as narrow band or wide band. Each has its advantages and disadvantages.

Just as with amplitude modulation the frequency domain has the center frequency, fc, and side frequencies spaced apart by the modulation frequency, fm. In the examples below I have chosen modulation indexes of 2.41 and 5.0. The modulation index of 2.41 has the property that the carrier frequency, fc, vanishes: does not even exist. Compare to AM that has three and only three component frequencies. In FM the number of component frequencies is infinite. However, those frequencies beyond fd get much smaller very quickly. So, a receiver designed to receive up to 2*fd will not miss much and will yield a good output.

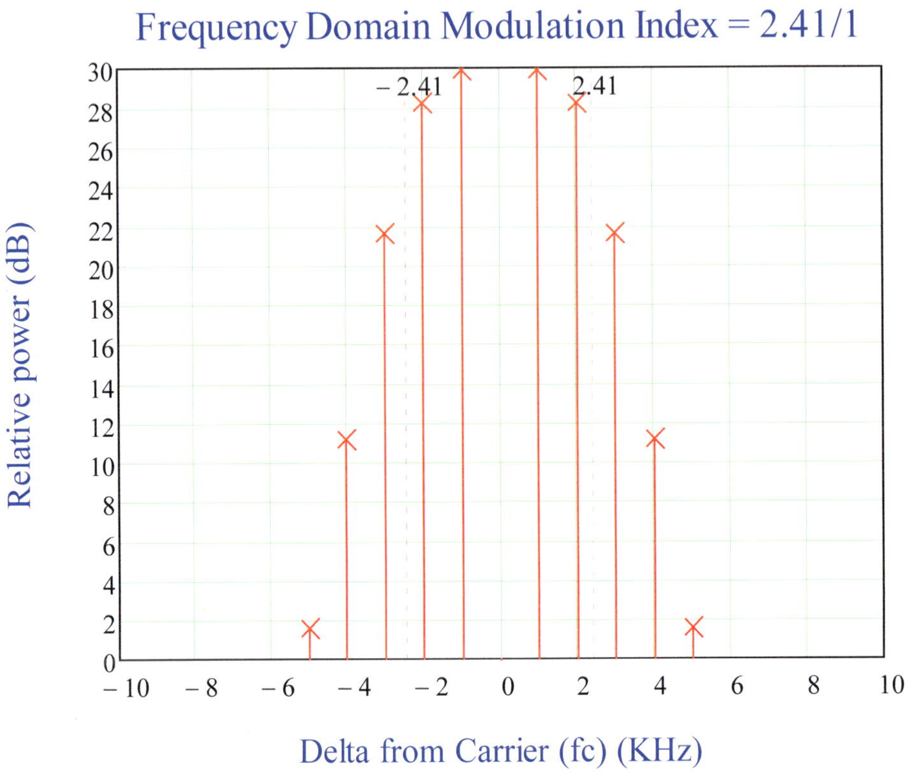

The mathematics to determine the level of each side tone is advanced, involving Bessel functions. Fortunately, there are tables of side tone levels to help us. Look up "FM Modulation" on Wikipedia to see an action graphic of how it works as well as the equations and the table.

Notice two things: First, there is zero power at the carrier frequency, fc.
Second, the power decreases very rapidly beyond the deviation frequency, fd.

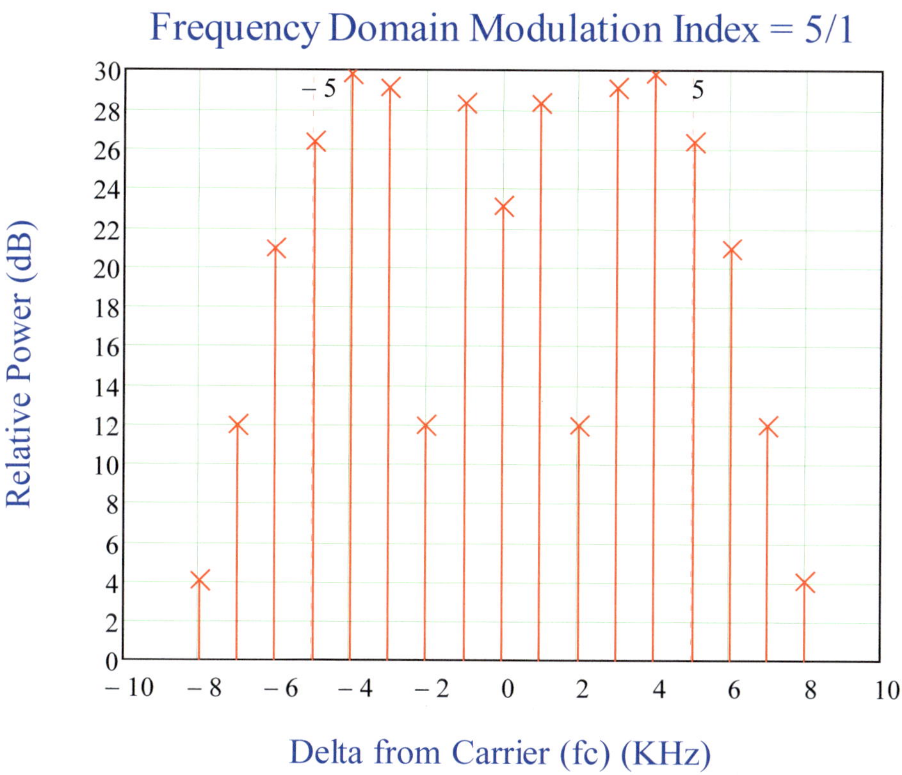

This time there are many more significant frequencies, but, as before, the levels decrease very rapidly beyond fd. There is no way from observation of the plots to determine the precise value of fd. (Note: the values of 0 to 30 on the y-axis have no meaning except to give a relative scale.) Also, not apparent from the plot, some of the delta frequencies are minus; i.e., 180 phase difference. The side tones do not stop as displayed in the plots. They extend to infinity, but are so low that they are insignificant and they are not needed to accurately demodulate the frequency, fm.

Harmonic Products of Non-Linearity

When a system such as an amplifier is non-linear, harmonically related signals are generated. Stated simply, if the input level is raised by 1 dB and the output level changes by more than or less than 1 dB, then the system is operating in a non-linear mode.

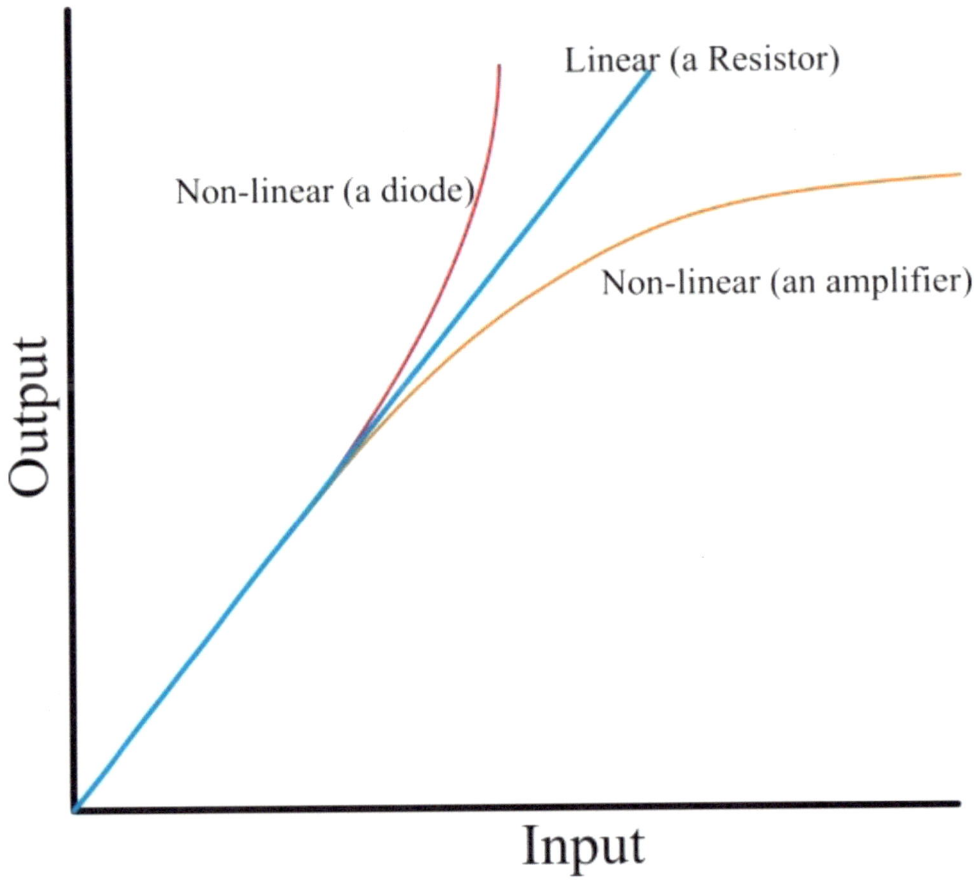

For this section, let's use the example of the saturating amplifier. At low input levels, the system is very nearly linear and time and frequency domain plots are:

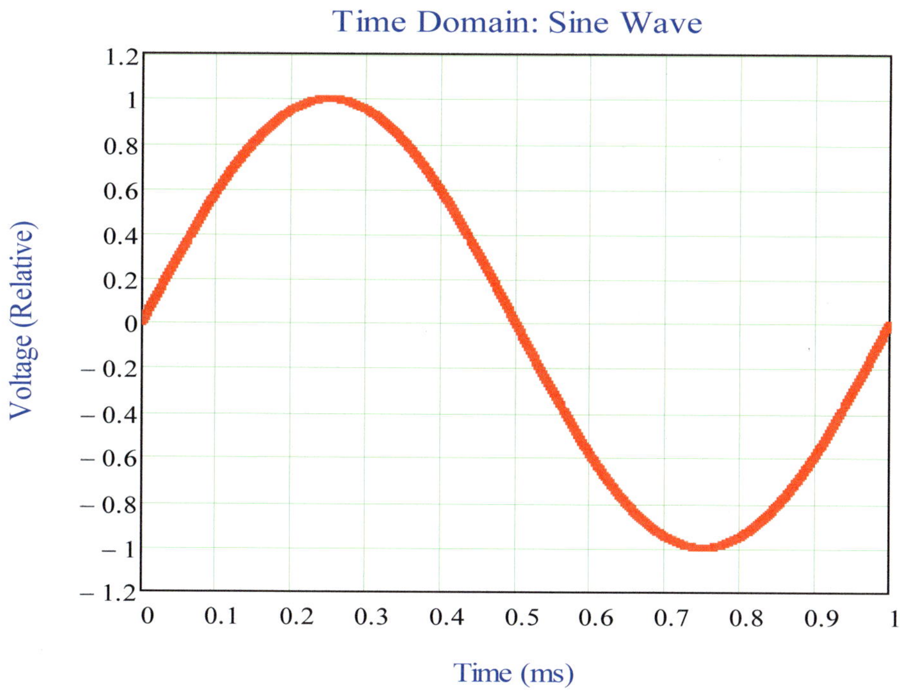
Time Domain: Sine Wave
Voltage (Relative)
1.2
1
0.8
0.6
0.4
0.2
0
– 0.2
– 0.4
– 0.6
– 0.8
– 1
– 1.2
0 0.1 0.2 0.3 0.4 0.5 0.6 0.7 0.8 0.9 1
Time (ms)

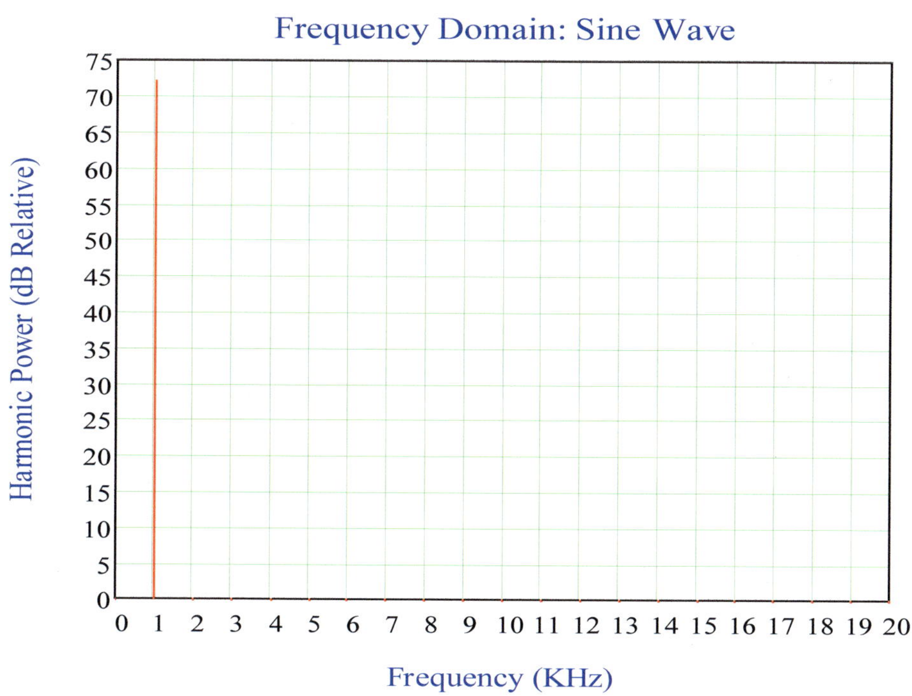
Frequency Domain: Sine Wave
Harmonic Power (dB Relative)
75
70
65
60
55
50
45
40
35
30
25
20
15
10
5
0
0 1 2 3 4 5 6 7 8 9 10 11 12 13 14 15 16 17 18 19 20
Frequency (KHz)

Output Distortion (Harmonics)

A typical non-linear system is an amplifier being driven into saturation. If you input a sine wave into a linear amplifier, you get out a sine wave. As the amplifier input is increased, the output begins to saturate; i.e. it clips the top of the output sine wave, giving it a "flat top"[35]. This generates new, harmonically related, sine waves. The following plots illustrate the emergence of the harmonics for certain degrees of clipping. The number in the title indicates the value of the sine wave where the clipping started, then amplified up to 1 v peak.

[35] In real life the shape is not as distinct as this, but this will demonstrate the effect.

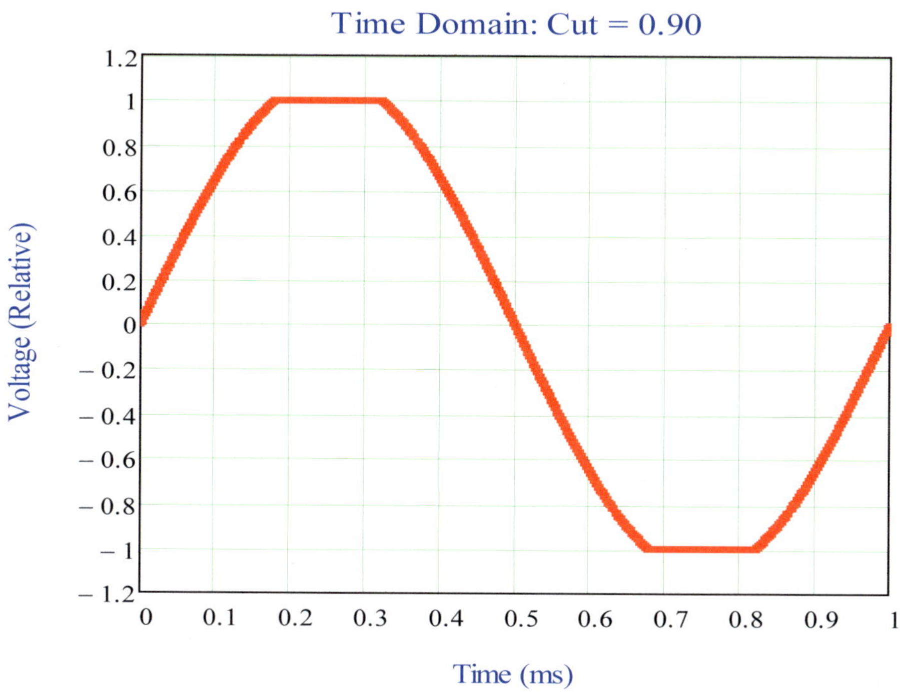
Time Domain: Cut = 0.90
Voltage (Relative)
1.2
1
0.8
0.6
0.4
0.2
0
– 0.2
– 0.4
– 0.6
– 0.8
– 1
– 1.2
0 0.1 0.2 0.3 0.4 0.5 0.6 0.7 0.8 0.9 1
Time (ms)

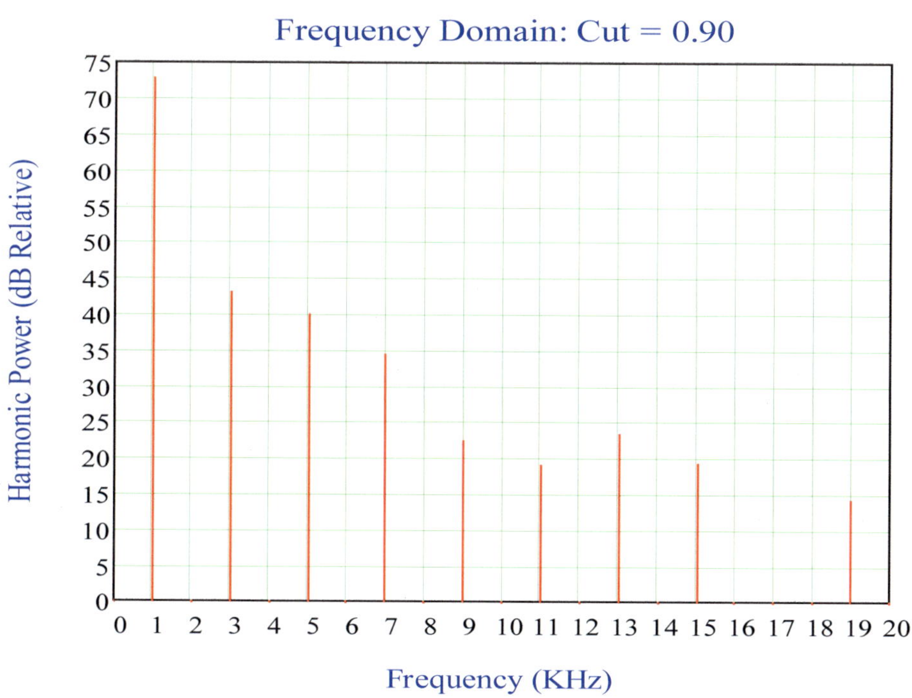
Frequency Domain: Cut = 0.90
Harmonic Power (dB Relative)
75
70
65
60
55
50
45
40
35
30
25
20
15
10
5
0
0 1 2 3 4 5 6 7 8 9 10 11 12 13 14 15 16 17 18 19 20
Frequency (KHz)

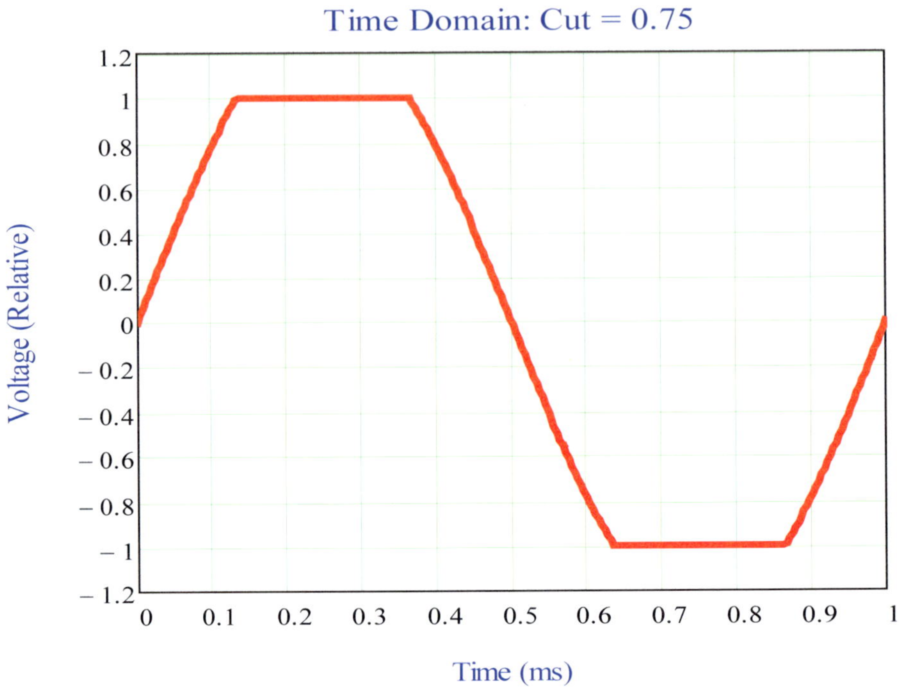
Time Domain: Cut = 0.75
Voltage (Relative)
Time (ms)

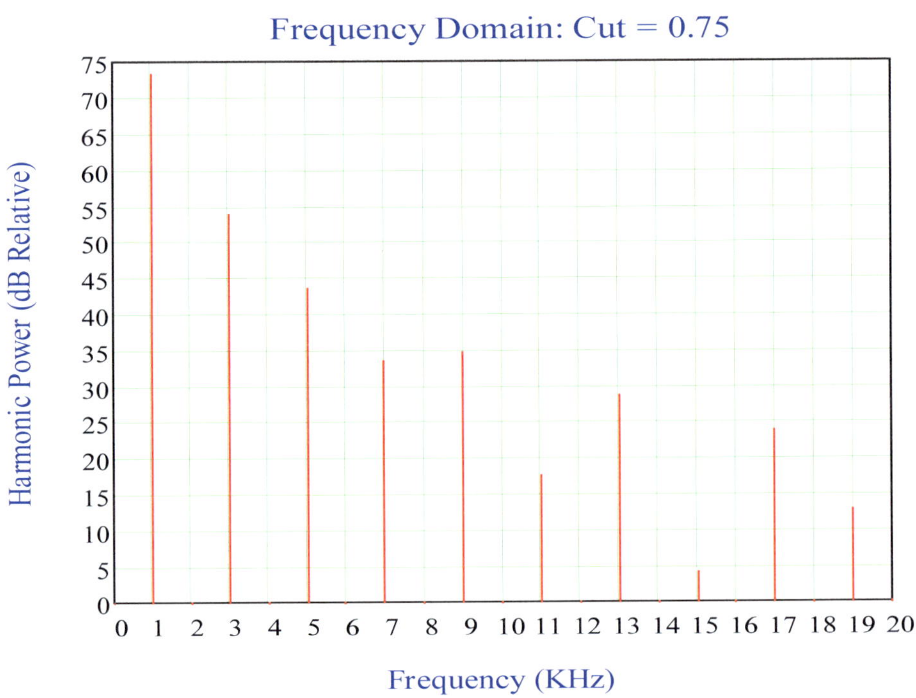
Frequency Domain: Cut = 0.75
Harmonic Power (dB Relative)
Frequency (KHz)

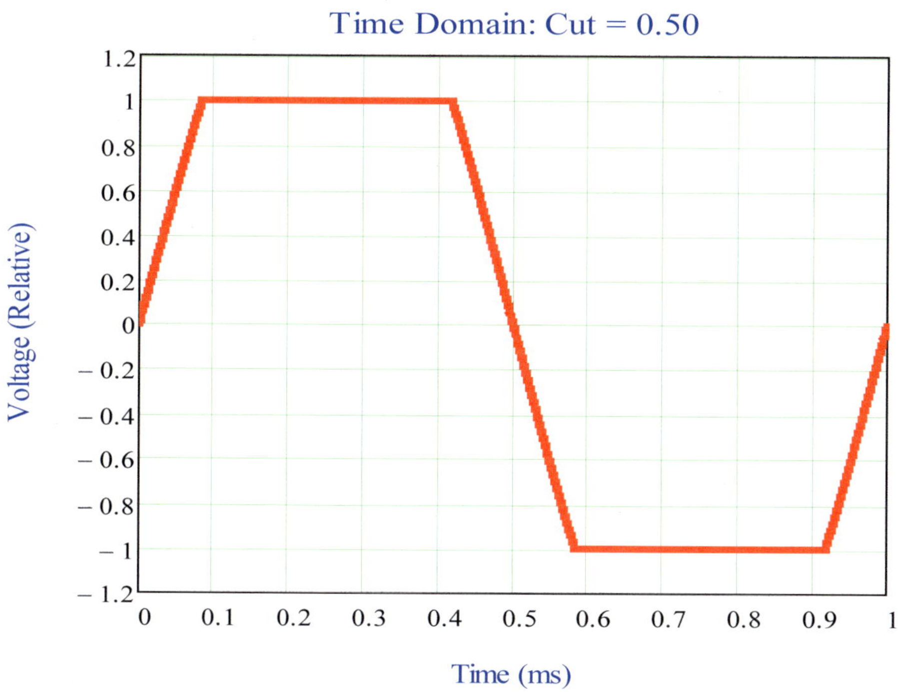
Time Domain: Cut = 0.50
Voltage (Relative)
1.2
1
0.8
0.6
0.4
0.2
0
– 0.2
– 0.4
– 0.6
– 0.8
– 1
– 1.2
0 0.1 0.2 0.3 0.4 0.5 0.6 0.7 0.8 0.9 1
Time (ms)

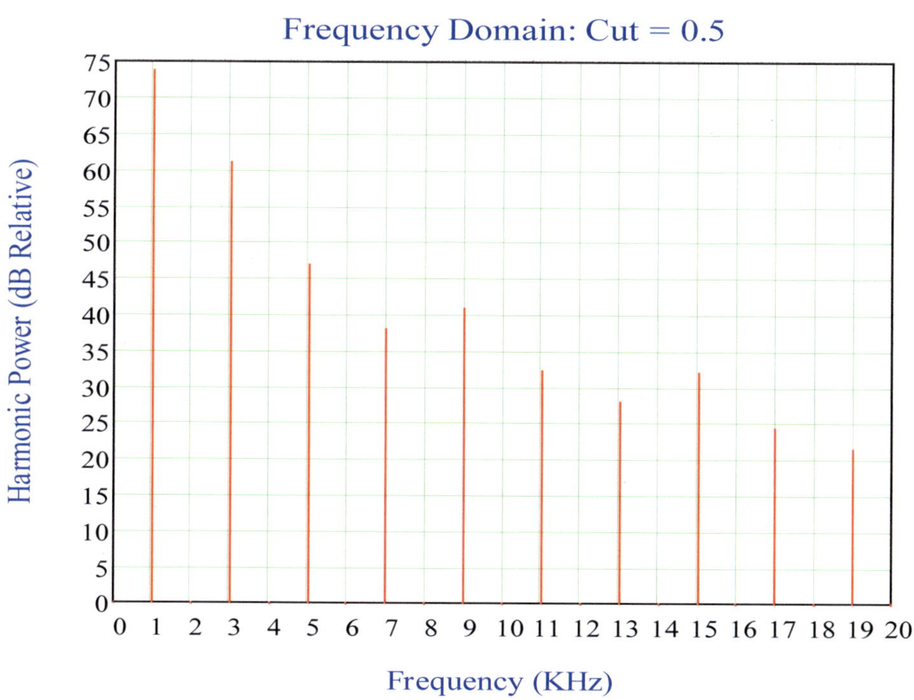
Frequency Domain: Cut = 0.5
Harmonic Power (dB Relative)
75
70
65
60
55
50
45
40
35
30
25
20
15
10
5
0
0 1 2 3 4 5 6 7 8 9 10 11 12 13 14 15 16 17 18 19 20
Frequency (KHz)

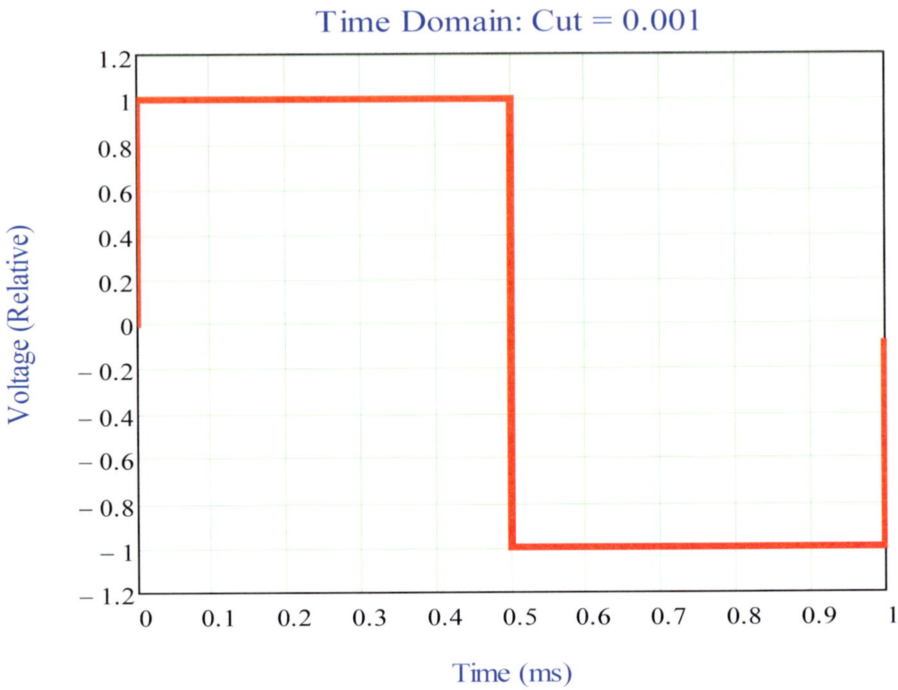
Time Domain: Cut = 0.001
Voltage (Relative)
Time (ms)
1.2
1
0.8
0.6
0.4
0.2
0
– 0.2
– 0.4
– 0.6
– 0.8
– 1
– 1.2
0 0.1 0.2 0.3 0.4 0.5 0.6 0.7 0.8 0.9 1

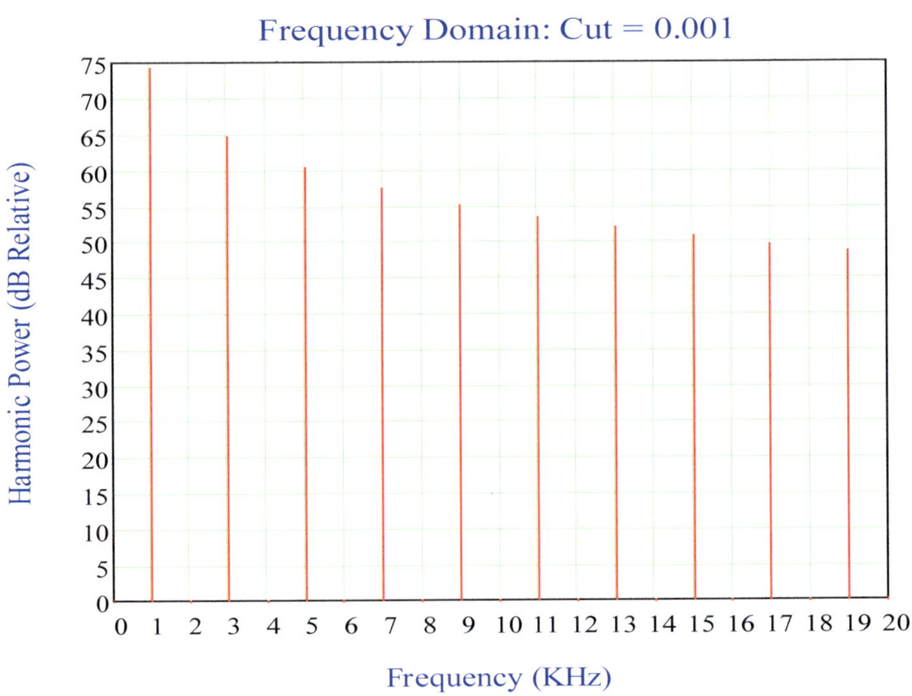
Frequency Domain: Cut = 0.001
Harmonic Power (dB Relative)
Frequency (KHz)
75 70 65 60 55 50 45 40 35 30 25 20 15 10 5 0
0 1 2 3 4 5 6 7 8 9 10 11 12 13 14 15 16 17 18 19 20

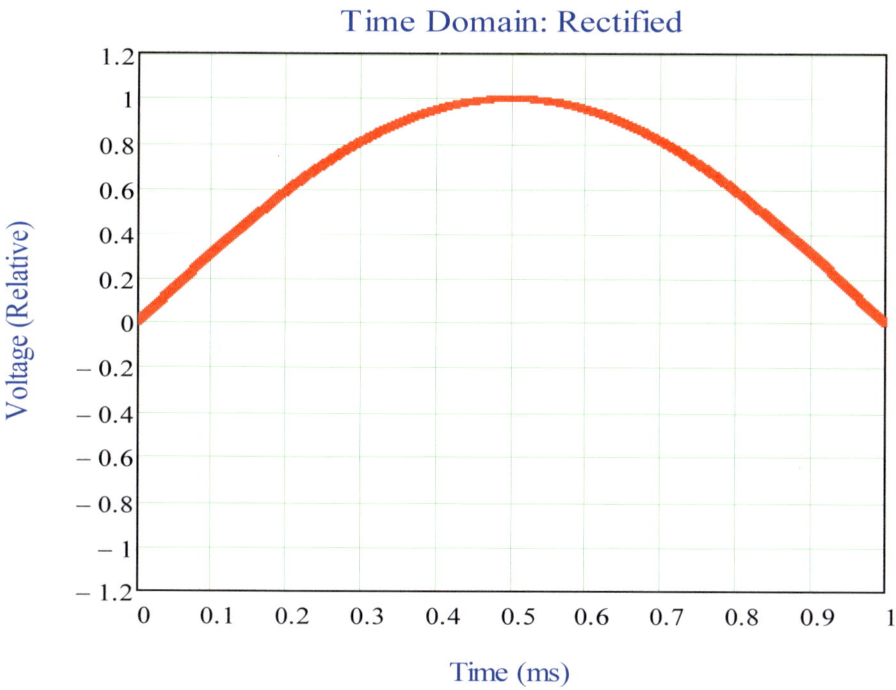
Time Domain: Rectified
Voltage (Relative)
1.2
1
0.8
0.6
0.4
0.2
0
– 0.2
– 0.4
– 0.6
– 0.8
– 1
– 1.2
0 0.1 0.2 0.3 0.4 0.5 0.6 0.7 0.8 0.9 1
Time (ms)

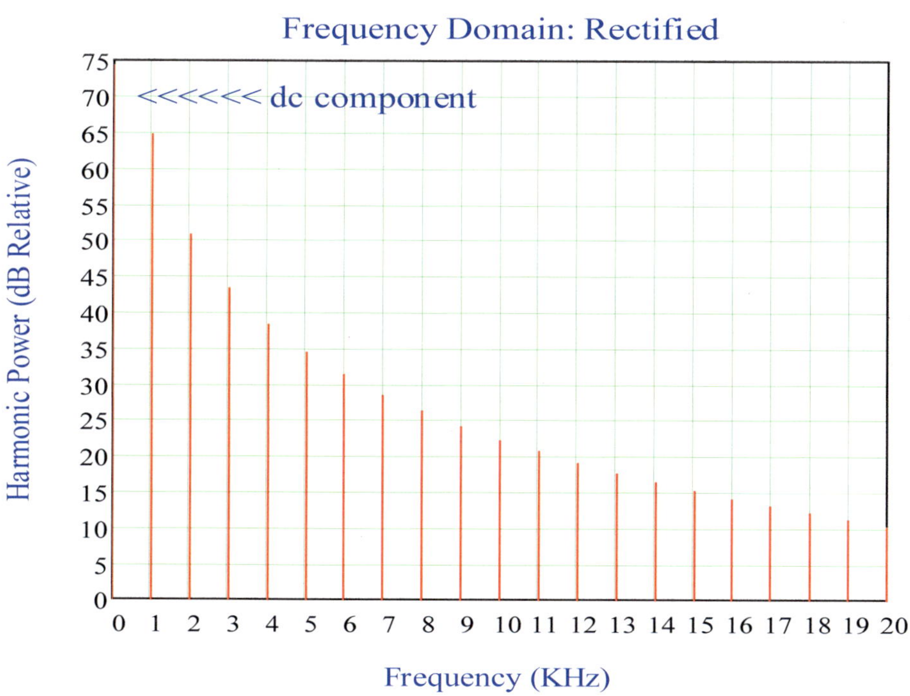
Frequency Domain: Rectified
Harmonic Power (dB Relative)
75
70
65
60
55
50
45
40
35
30
25
20
15
10
5
0
<<<<<< dc component
0 1 2 3 4 5 6 7 8 9 10 11 12 13 14 15 16 17 18 19 20
Frequency (KHz)

Notice the dc component indicating that the waveform is not symmetric around zero.

These are the program equations used to generate the plots. The frequency domain plot is the Fourier transform of the time domain plots. There are many applications in EM analysis where the Fourier transform is used.

Time Domain Program

$T \equiv 1 \cdot ms \qquad F \equiv \frac{1}{T} \qquad cut \equiv 0.001$

$$
\begin{aligned}
Sum2000 \equiv \; & \text{for } i \in 0 .. 2^m - 1 \\
& \quad t_i \leftarrow \left[\frac{i}{1000 \cdot \left(2^m\right)} + 10^{-10}\right] \cdot s \\
& \quad Y_i \leftarrow \sin\left(2\pi \cdot F \cdot t_i\right) \\
& \quad Y_i \leftarrow cut \;\; \text{if } Y_i > cut \\
& \quad Y_i \leftarrow -cut \;\; \text{if } Y_i < -cut \\
& \quad 0 \\
& \text{augment}\left(t, \frac{1}{cut} \cdot Y \cdot s\right)
\end{aligned}
$$

Frequency Domain Program (Fourier Transform)

$$m \equiv 16$$

$$fs \equiv 2^{m}$$

$$fs = 6.554 \times 10^{4}$$

$$k \equiv 0..2^{(m)}$$

$$f_k \equiv \frac{k}{2^m} \cdot fs$$

$$Amp \equiv \overrightarrow{\left|\mathrm{fft}\left(\left(Sum2000^{\langle 1\rangle}\right)\right)\right|} \div s$$

Replace the 0 with this for rectified.

$$Y_i \leftarrow (-Y)_i \;\; \text{if} \;\; Y_i < 0$$

Another way to demonstrate the effect of the harmonics is to use sine raised to a power.

$$\sin X = \sin X$$

$$\sin^2 X = 1/2 - (1/2)\cos(2X))$$

$$\sin^3 X = (3/4)\sin X - (1/4)\sin(3X)$$

$$\sin^4 X = (3/8) - (1/2)\cos(2X) + (1/8)\cos(4X)$$

$$\sin^5 X = (5/8)\sin X - (5/16)\sin(3X) + (1/16)\sin(5X)$$

$$\sin^6 X = 5/16 - (15/32)\cos(2X) + (6/32)\cos(4X) - (1/32)\cos(6X)$$

Observe that each one generates integer harmonics.

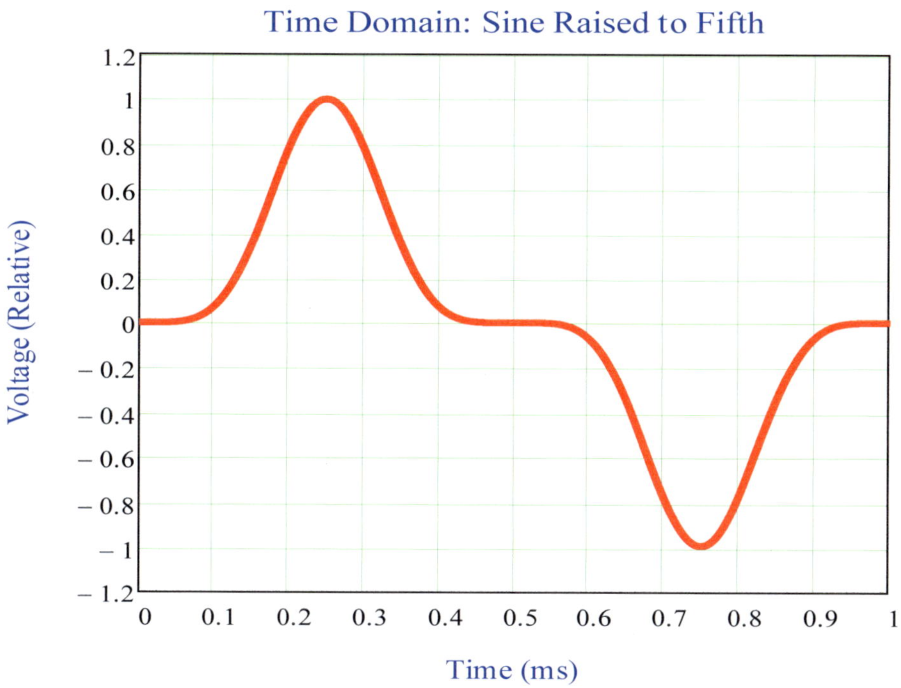
Time Domain: Sine Raised to Fifth
Voltage (Relative)
1.2
1
0.8
0.6
0.4
0.2
0
– 0.2
– 0.4
– 0.6
– 0.8
– 1
– 1.2
0 0.1 0.2 0.3 0.4 0.5 0.6 0.7 0.8 0.9 1
Time (ms)

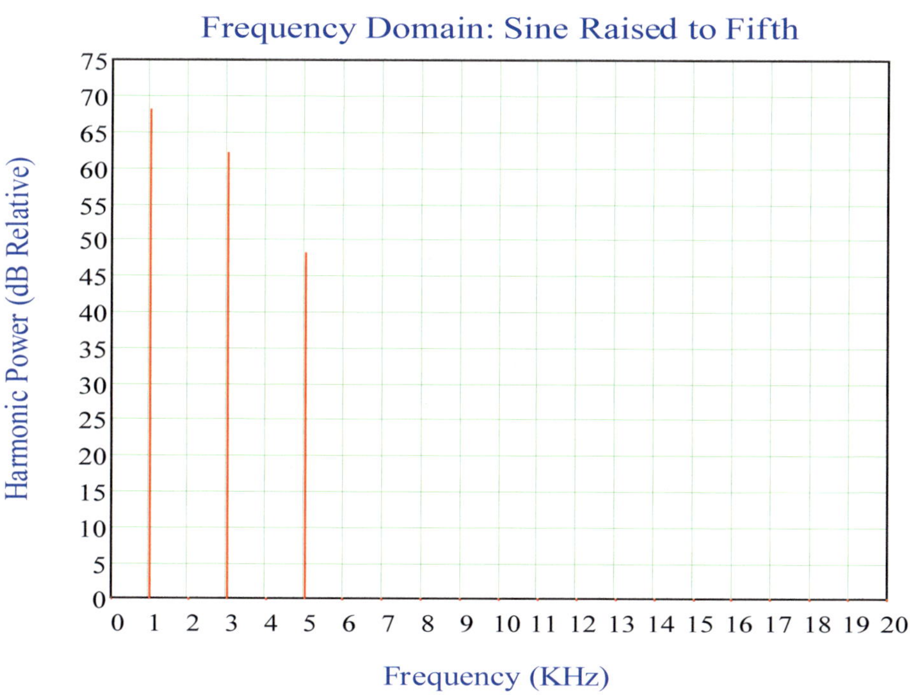
Frequency Domain: Sine Raised to Fifth
Harmonic Power (dB Relative)
75
70
65
60
55
50
45
40
35
30
25
20
15
10
5
0
0 1 2 3 4 5 6 7 8 9 10 11 12 13 14 15 16 17 18 19 20
Frequency (KHz)

This time, in accordance with above chart, only three harmonics are generated. Observe that the time domain plot is in volts and the frequency domain plot is relative power, 20∗log(volt).

Small and Large Antennas

I mentioned previously that the size of objects relative to the wave length of the signal is important. Nowhere is that more important than antennas. In fact you can classify antennas in two categories: Smaller than one wavelength and larger than one wavelength. So, when we reference the size of an antenna, we do not mean its physical size, but its wavelength size.

Examples of small antennas are Yagi, Log Periodic, Dipole, Monopole, Patch, Spiral, etc. Large antennas include Helix, Parabolic Dish, Array of Elements. Some antennas are very large indeed, approaching 100 wavelengths. Here is physical law to remember. **"The beamwidth of an antenna pattern is inversely proportional to its size."** Large antennas have a narrow beamwidth. That means that a large receiving antenna must be pointing precisely toward the source of the radiation.

So, the very large astronomy antennas out in the desert are "looking at" a very small portion of the sky.

In the following sections, I will be discussing large antennas, both as a single structure and as an array of small antennas, termed its elements.

Antenna Pattern

I have used the term "antenna pattern" without defining it. So, let's do that now. I will use a large antenna because the presentation is more graphic. The antenna aperture is the physical extent of the antenna. In a horn antenna it is the area of the face of the horn. In a parabolic reflector antenna it is the area of the face of the parabolic dish. In the section on arrays we will find that is it the extent of the face of the array.

A characteristic of an aperture antenna is that the amplitude and phase of the radiation[36] is the same (or nearly so) over the whole space of the aperture. For this demonstration assume that is so.

The term "antenna pattern" or "radiation pattern" means the distribution in space of the amplitude (voltage) of the radiation emitted from the antenna aperture. The reference is the perpendicular to the antenna aperture. An observer (receiver) in the far-field sees a maximum voltage when located on the line perpendicular to center of the aperture plane. As the observer moves laterally away from the perpendicular, the voltage level decreases. For very large apertures, this happens very quickly because of the narrow beamwidth. Beamwidth is defined as the extent between the points where the radiation has decreased to 0.707 times the peak voltage. When normalized, power is voltage squared so the power is $0.707^2 = 0.5$ (half power)[37]

---

[36] In passive antennas (no amplifier) transmission and reception are reciprocal; meaning characteristics associated with transmission are true also for reception. Usually it is easier to describe an effect by speaking of transmission. I do this even when by my subject antenna is a receive only type.

[37] The relies on the assumption that half of the power radiated is between the 3 dB points and the other half of the power lies in all of the remaining space. This is not always precisely true, but is close enough that it has become the industry standard for defining beamwidth.

Look at this graphic:

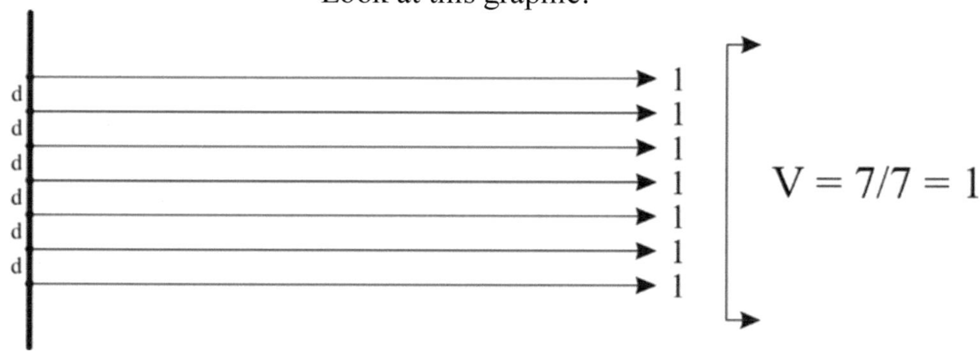

On a linear antenna seven points are pointing at boresite in the far-field. The distance is so great that all of the directions are essentially parallel. The voltage in the far-field is 7 which is normalized by seven to yield 1.0.

Now look from an angle (θ) off boresite:
Each ray now travels a different distance to the far-field point. The distance difference is n∗d∗sin(θ) where n is the number of spaces from center. The distance difference introduces a phase lead or lag. Now Euler's equation once again comes to the rescue and allows us to add them all together as shown here.

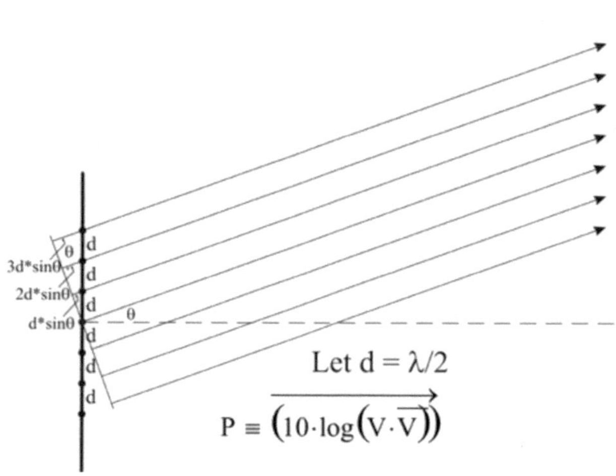

$$V \equiv \left[\left[\frac{\sum_{n=-3}^{3} \left(e^{j\cdot 2\cdot\pi\cdot n\cdot d\cdot \sin(\theta)}\right)}{7}\right]\right]$$

$$P \equiv \overrightarrow{\left(10\cdot\log\left(V\cdot\overline{V}\right)\right)}$$

Here is the plot of radiation power versus look angle:

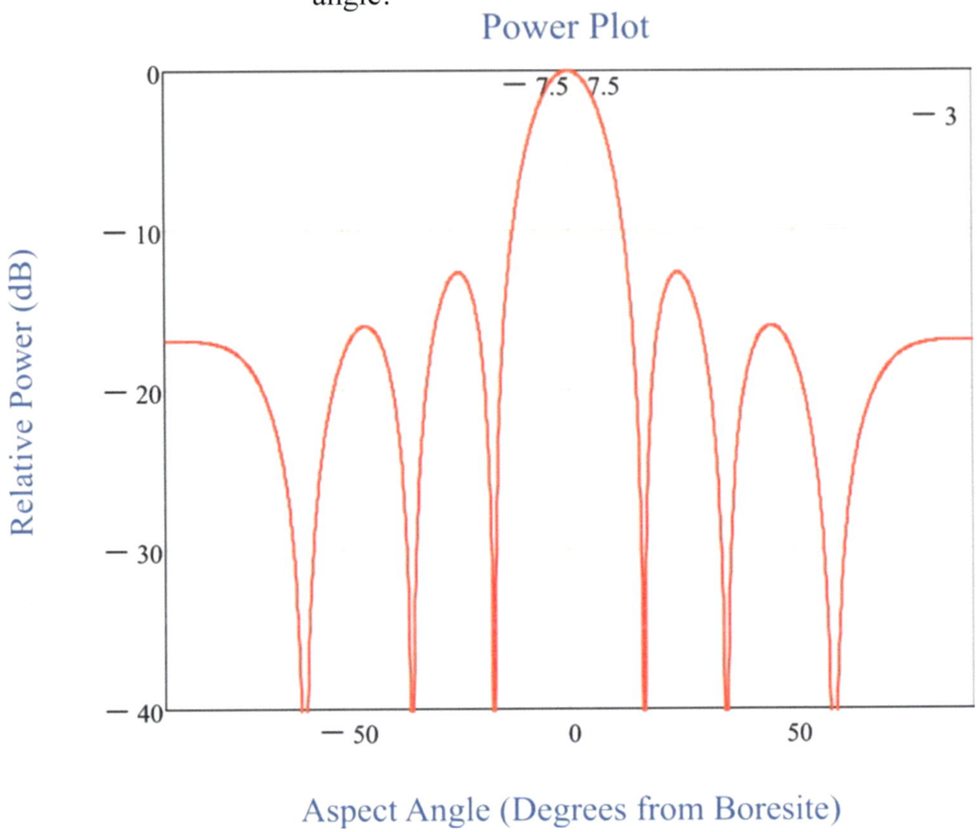

The three dB down points are at ±7.5 degrees, so, the beamwidth is 15 degrees.

It is a characteristic of large aperture antennas to have a main lobe and many side lobes. The condition where all radiating points have the same amplitude and phase is called uniform distribution. In this case the first side lobe is 13.6 dB below the main peak. By adjusting the phase of the radiation points, the level of the side lobes can be controlled to be different values, but the main beamwidth widens. **Rule: Uniform distribution yields minimum beamwidth and maximum directive gain**. More on this in the section on arrays.

EM Striking an Object

When a travelling EM wave encounters an object, two things happen. Part of the wave is reflected back and part of it transmits into the object. In the special case where the object is a very good conductor such as copper, the entire wave is reflected back.

Study this graphic.

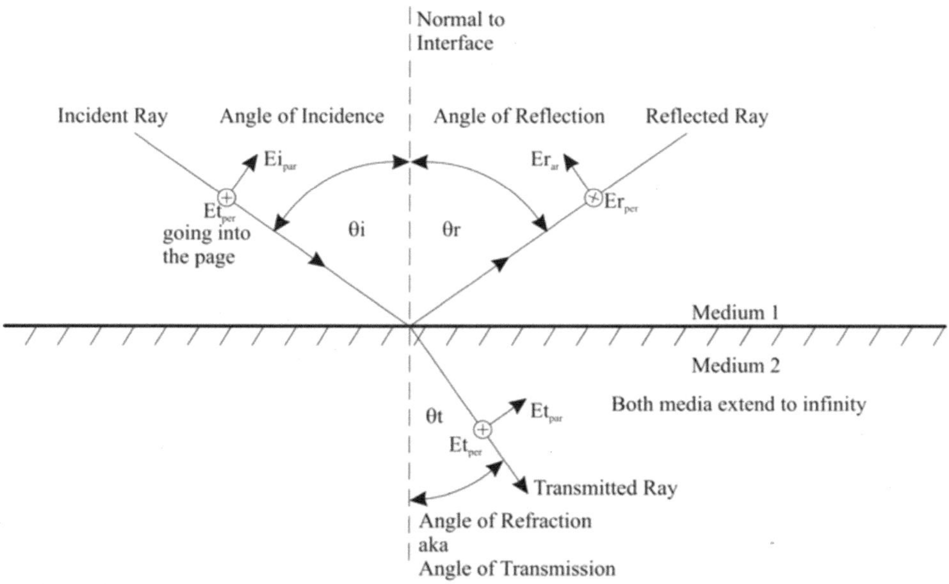

Plane of Indecence (POI) contains the Incident Ray, the Reflected Ray, the Transmitted Ray and the Normal
Define two E vectors, one perpendicular to the POI and the other parallel to the POI. (The parallel
E vector lies in the POI. Both E vectors are perpendicular to the Incident Ray.

An EM wave polarization vector can have any orientation in space, but recall that any vector can be resolved into two (or more) vectors. We make use of that here and resolve our vector into a parallel component and perpendicular component as described above because there are equations for these two cases. There are two things we want to know. How much of the wave was reflected (reflection coefficient) and how much was transmitted into medium 2 (transmission coefficient).

The permittivity of free space is $8.854*10^{-12}$ Farads/meter. The permeability of free space is $1.257*10^{-6}$ Henry/meter.

The permeability comes into play only when either medium is magnetic, which is almost never the case.

If we further assume that both media are lossless (no heating), then we only need the permittivity relative (εr) to free space. (Also known as "dielectric constant".) Free space is 1.0.

It was shown earlier that the impedance of free space is 377 ohms. The impedance of other materials is 377 divided by the square root of the relative dielectric constant of material.

Given these assumptions, we need only the relative dielectric constant to solve our equations.

If you decide to pursue this further to include losses, there are many fine books to show you how to do that. But, a word of caution. The authors go about the solution of the equations very differently. It is hard to compare them, but they eventually come to the same answer.

First define the media. Let medium 1 be air, εr = 1 and medium 2 be earth, εr = 3.5.

The "index of refraction" is the ratio of the velocity of propagation in vacuum to the velocity of propagation in the material. The dielectric constant is the square root of the index of refraction.

Here are the equations:

$Zo \equiv 377$

Air

$\varepsilon1 \equiv 1$

$Z1 \equiv \frac{Zo}{\sqrt{\varepsilon1}} = 377$

Earth

$\varepsilon2 \equiv 3.5$

$Z2 \equiv \frac{Zo}{\sqrt{\varepsilon2}} = 201.5$

Let

$\theta i \equiv 61.9 \cdot deg$

$\theta r \equiv \theta i = 61.9\,deg$

$$\theta t \equiv asin\left(\sqrt{\frac{\varepsilon1}{\varepsilon2}} \cdot sin(\theta i)\right) = 28.1\,deg$$

$$\rho_{par}(\theta i) \equiv \frac{-\left(\frac{\varepsilon2}{\varepsilon1}\right) \cdot cos(\theta i) + \sqrt{\left(\frac{\varepsilon2}{\varepsilon1}\right) - sin(\theta i)^2}}{\left(\frac{\varepsilon2}{\varepsilon1}\right) \cdot cos(\theta i) + \sqrt{\left(\frac{\varepsilon2}{\varepsilon1}\right) - sin(\theta i)^2}} \qquad \rho_{par}(\theta i) = 0$$

$$\rho_{per}(\theta i) \equiv \frac{cos(\theta i) - \sqrt{\left(\frac{\varepsilon2}{\varepsilon1}\right) - sin(\theta i)^2}}{cos(\theta i) + \sqrt{\left(\frac{\varepsilon2}{\varepsilon1}\right) - sin(\theta i)^2}} \qquad \rho_{per}(\theta i) = -0.556$$

Here is a plot from normal (0 degrees) to 90 degrees incident angle.

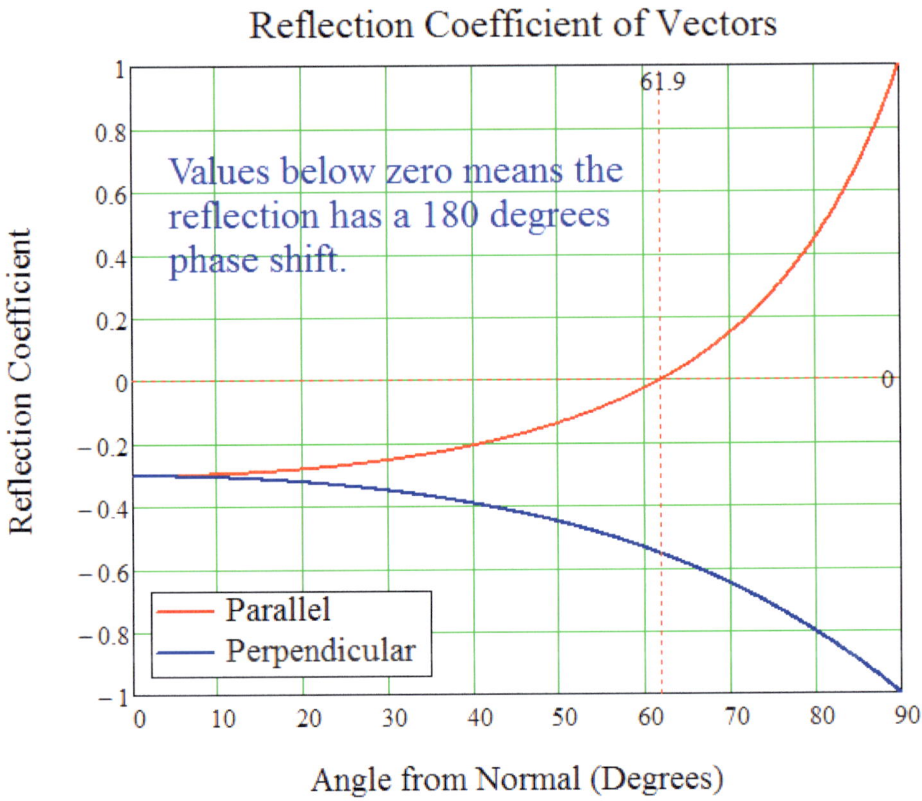

Observe that at normal (0 degrees) both polarizations are the same and the reflection induces a 180 degrees phase shift. Now look at incident angle of 61.9 degrees. The reflection coefficient is zero. That means that all of the energy was transmitted into medium 2, none reflected. Higher than 61.9 degrees the reflection has zero degrees phase shift. I have been nice and show the graph the way it should be plotted, but most texts give you this.

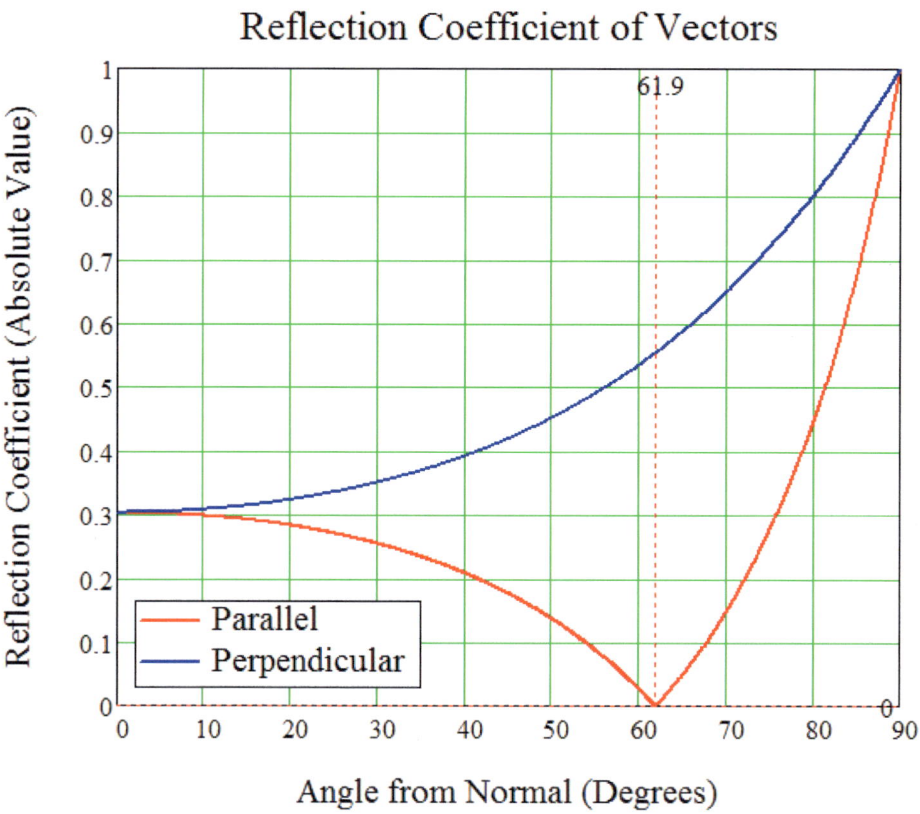

Thus hiding the effect of the 180 degrees phase shift. Some texts will have an accompanying graph showing phase, but many do not. I think it is lazy of them!

Here is the transmission coefficient. (The part that is going into medium 2.)

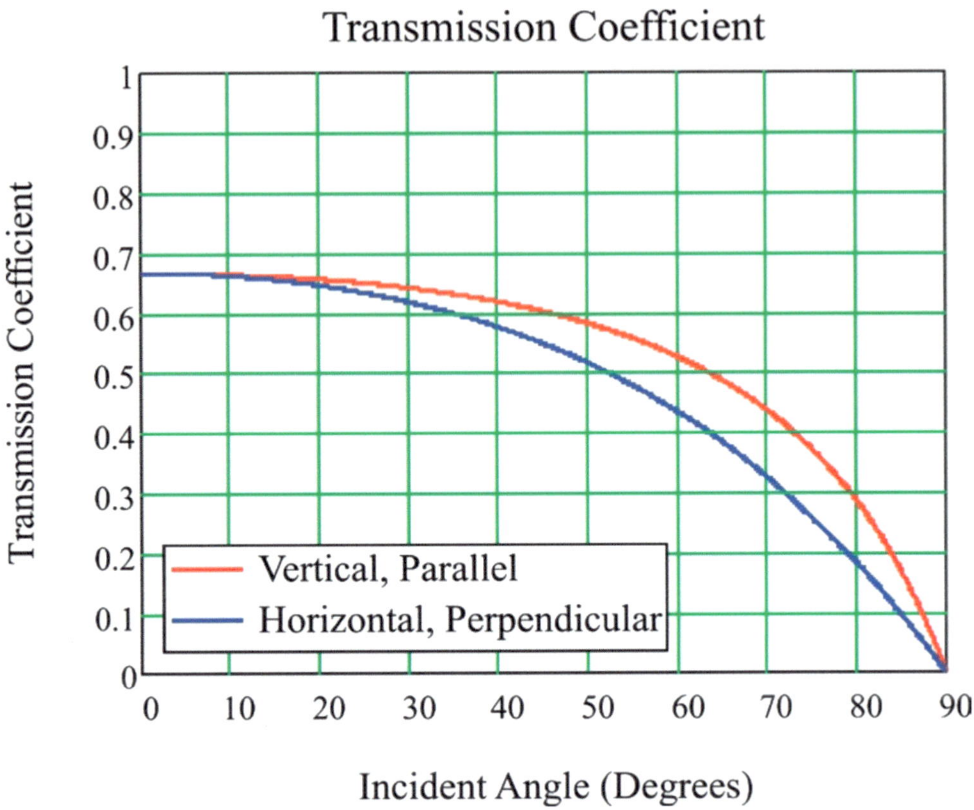
Transmission Coefficient
Transmission Coefficient
1
0.9
0.8
0.7
0.6
0.5
0.4
0.3
0.2
0.1
0
0
10
20
30
40
50
60
70
80
90
Vertical, Parallel
Horizontal, Perpendicular
Incident Angle (Degrees)

Directivity and Area

Let me restate here the meaning of directivity, aka directive gain. Locate an antenna inside a sphere. If the antenna radiates the same power in every direction, then every place on the sphere has the same power density (Watts per square meter.) This is the impossible condition is known as the omni-directional antenna. Although the omni antenna is impossible in real life, it is none the less used as standard of measurement. Any real antenna has higher power density on some part of the sphere and lower power density on another part of the sphere. The magnitude of the highest density point divided by the magnitude for the "omni" value is its directivity (directive gain). When expressed in logarithms (dB) the directive gain is X dBi, meaning X dB above isotropic. Isotropic is 0 dBi [38].

If an antenna has heat producing loss, the output power is less, but the directive gain remains unchanged. The equation is: Antenna Gain = Antenna Directive Gain – Antenna Loss. Most antennas have such low loss that the terms are interchangeable, but not always, so watch for it.

Antenna Area

One equation for antenna gain is Gain = $4\pi A_e/\lambda^2$, where A_e is the effective area, not the physical area, of the antenna. In my mind, it is not a very useful formula except in this form: A_e = Gain$*\lambda^2/4\pi$. As we will see later, this equation is useful in computation of the link budget. Of course it assumes that we know the gain, probably from manufacturer's data or direct measurement.

---

[38] Remember that in logarithms, division becomes subtraction.

This gain can, and probably does, include any heating loss.

The physical area of a half wave thin wire dipole is nearly zero, but the effective area is $0.1305\lambda^2$; that is very nearly the size of a rectangle ½ λ by ¼ $\lambda = 0.125\lambda^2$ So, this very thin wire literally "sucks in" a large area of energy near it, emphasizing the disconnect between effective area and physical area.

Intercept Area

Half Wave Dipole

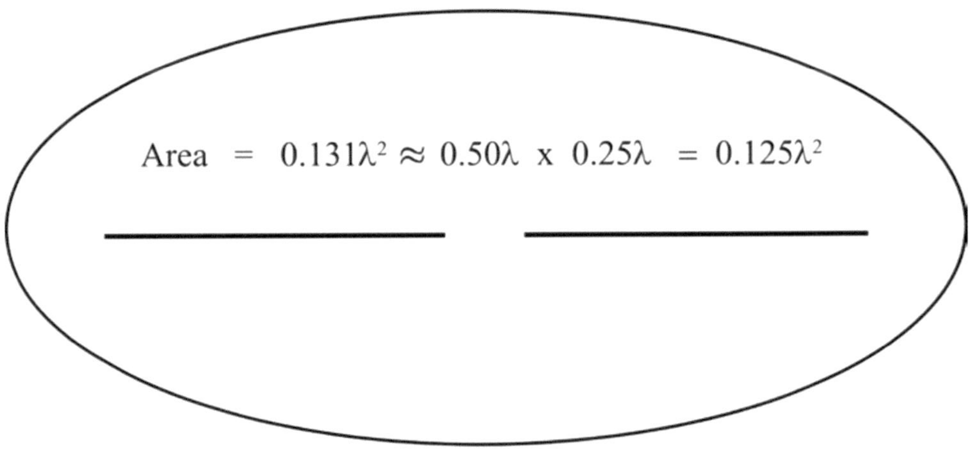

At the other end of the scale, the opposite is true. The effective area of a parabolic reflector is less than its physical area.

Antennas that comes closest to a match include a large array where all of the array elements are driven with the same power level and a large, flat conducting sheet receiving a wave from very far away..

Even the isotropic antenna has an effective area. It is $\lambda^2/4\pi = 0.0796\lambda^2$.

Directive Gain and Beamwidth

This section discusses the relationship of directive gain and beamwidth; or, more precisely, between directive gain and the antenna pattern. Rule: **The directive gain of an antenna is <u>solely</u> a function of the shape of the antenna pattern.**[39] That includes all the power density passing through the surface of the sphere enclosing the antenna. The only way to precisely determine the directive gain mathematically is to integrate the power density over the sphere and relate the peak value to the omni case.

Let's start by looking at the geometry of the sphere:

---

[39] I have found that many people have a hard time accepting this fact. They tend to attribute other characteristics to the determination of directive gain.

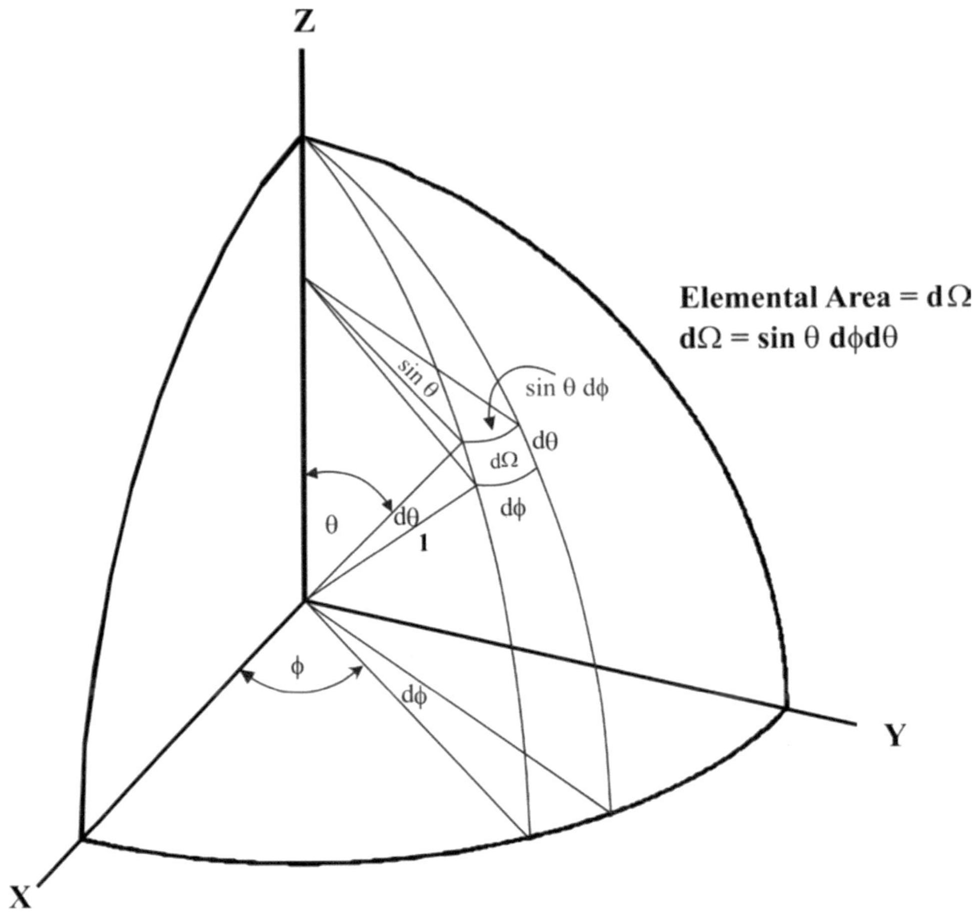

Coordinate System

The integration process starts by creating the curved rectangle dΩ, that sets the area through which the power density passes. Power density (Watts/meter^2) times the area equals the power for that point. For mathematical integration, dΩ becomes infinitely small. For piecewise integration, the area is set as small a practical for the computation software. To be rigorously correct, you must integrate the entire sphere, but, in practice, the areas where the radiated power is very low need not be included.

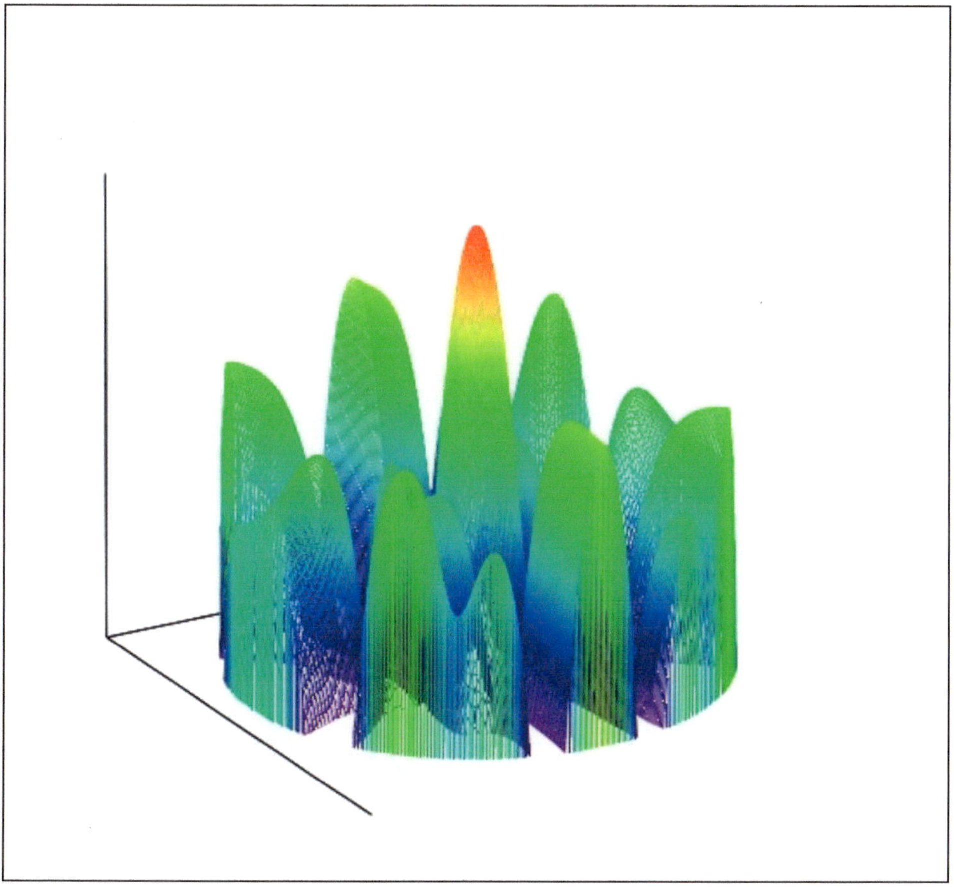

GU

Here is an example of a radiation pattern in 3-D. Only the upper hemisphere is shown. This next plot shows the "cut" in two orthogonal positions, commonly azimuth and elevation.

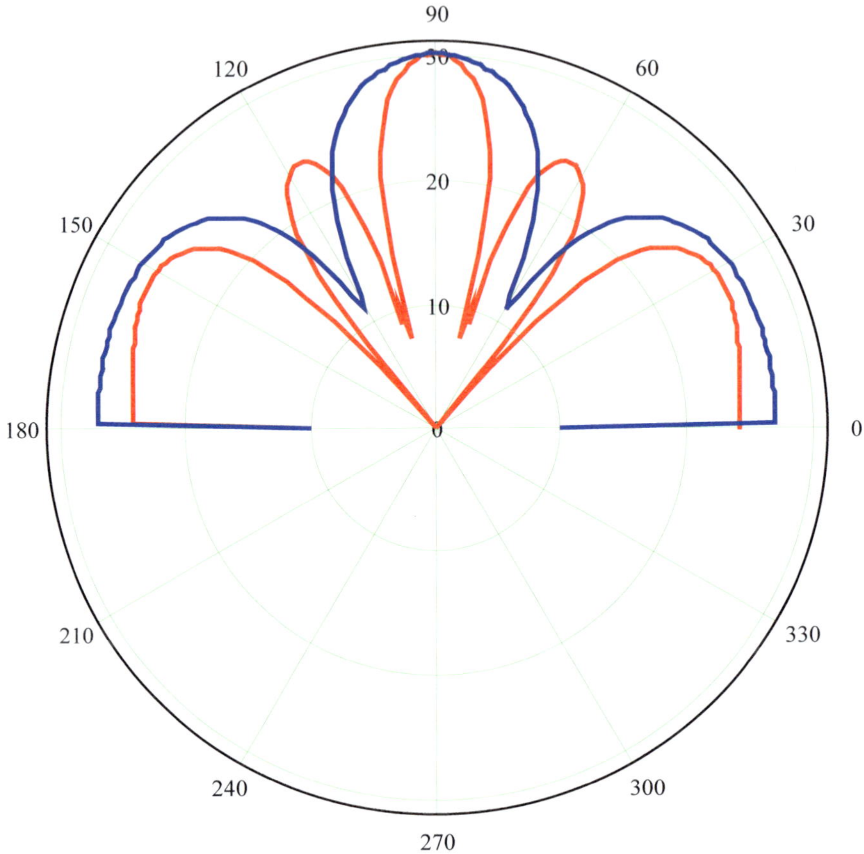

The beamwidth of the two patterns are commonly used to estimate the directive gain.

The exact relationship of directivity to antenna pattern is:

$$\text{Directivity} = \frac{4\pi}{\int\int U(\theta,\phi)\sin\theta \, d\theta d\phi} \quad \text{(Normalized}$$

Where U(θ,ϕ) is the radiated power density in the direction of θ and ϕ. The normalization means that U(θ,ϕ) has been scaled such that the maximum value of U(θ,ϕ) is 1.0.

When the values associated with θ are independent of the values associated with ϕ, then the equation can be separated.

$$D = \frac{4\pi}{\iint U(\theta,\phi)\sin\theta \, d\theta d\phi} = \frac{2\pi}{\int_0^{2\pi} U(\phi)\, d\phi} \times \frac{2}{\int_0^{\pi} U(\theta)\sin\theta \, d\theta}$$

Remembering that half of the power is between the 3 dB points, observe this figure.

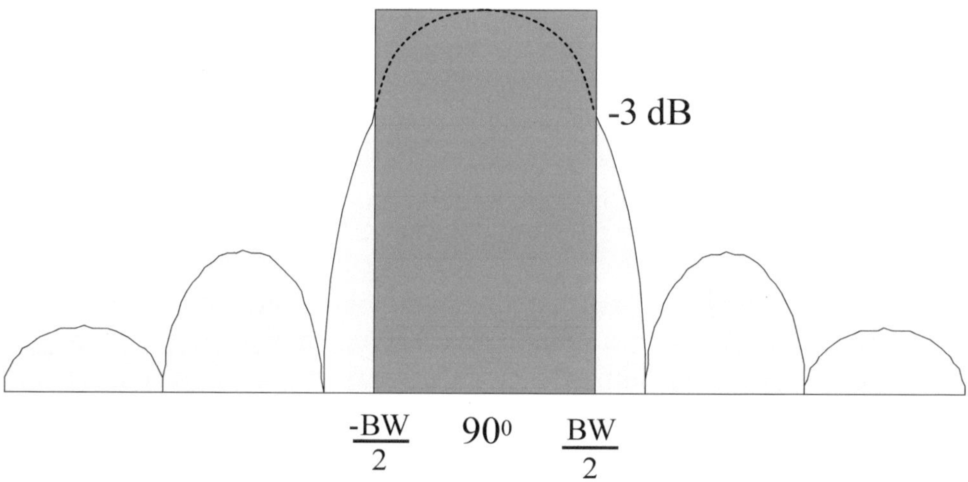

and:

$$U(\phi) = 1 \qquad U(\theta) = 1$$

$$\text{Directivity} = \frac{2\pi}{\int_{90-BW/2}^{90+BW/2} d\phi} \times \frac{2}{\int_{90-BW/2}^{90+BW/2} \sin\theta \, d\theta}$$

This is a new equation that I invented and copyrighted.[40] I like it for its simplicity (easy to

[40] Copyright number TXu 00180792

remember). Stated in a sentence – **"Directivity is 2π (360) divided by the phi beamwidth times the sine of half of the theta beamwidth."** Interchanging the two beamwidths yields essentially the same answer. This equation is more accurate than others in the literature.

$$\text{Directivity} = \frac{2\pi}{BW\phi \ \sin(BW\theta/2)} \quad \text{or} \quad \frac{360}{BW\phi \ \sin(BW\theta/2)}$$

(in radians) (in degrees)

There is only one practical way to precisely ascertain the directivity of an antenna. You must integrate measured values over the surface of the sphere. This requires a lot of measurements closely spaced on the sphere. It can take hours to gather enough points. It is not worth it unless you are trying to determine the loss of the antenna.

The usually methodology of measuring antennas is to calibrate the test range with a standard gain horn (SGH) and compare the antenna under test (AUT) to the SGH.[41] The value from the AUT minus the gain of the SGH is the gain of the AUT. But this yields physical gain, not directive gain. In other words, the loss of the AUT is obscured.

I will say it again: **The directive gain of an antenna is determined <u>solely</u> the shape of its radiation pattern.** If the beamwidth changes, the directivity changes!! This is an effect that many in the EM community have a hard time wrapping their minds around. Remember that real physical gain is directive gain minus antenna loss (dB).

---

[41] The standard gain antennas are calibrated to mathematical equations by a standards laboratory.

Signal Projected from Antenna and Far Field

Here is something I will bet you have not seen before.

One of the questions that I am most often asked is, "What is the far-field?"[42]. The answer is, "It depends.". It depends on what are the most important requirements for your antenna, gain, beamwidth, phase, etc. Many people have the misconception that if the antenna is not in the far-field that it will not work. That is not true, but it does raise the question, "What is the anternna radiation pattern if the antenna is closer than the "far-field". Well, I am going to show you. This demonstration works better for large antennas with narrow beamwidth. So, in this figure, the Antenna Aperture Plane can be the surface across the face of a parabolic reflector antenna. Then the phase at every point is the same. The amplitide will roll off toward the edge.

At some distance away in space, create an imaginary line with secondary radiation points. Then, separately sum every point on the Antenna Aperture Plane to every point on the secondary plane. The seconday plane must be wider because of spreading of the wave. Every point on the seconday plane will have a new amplitude and phase and the new phases are no longer equal. But summing the secondary aperture to the far-field yields the same answer as projecting from the antenna itself.

Now, we can get the antenna radiation pattern shape at any distance, R, from the antenna.

---

[42] The conventional definition of the far-field of a large antenna is $2D^2/\lambda$ where D is the greatest dimension of either the transmit or the receive antenna.

I chose the frequency of 11.8 GHz because that makes the wavelength one inch. The size of the dish is D = 6 feet.

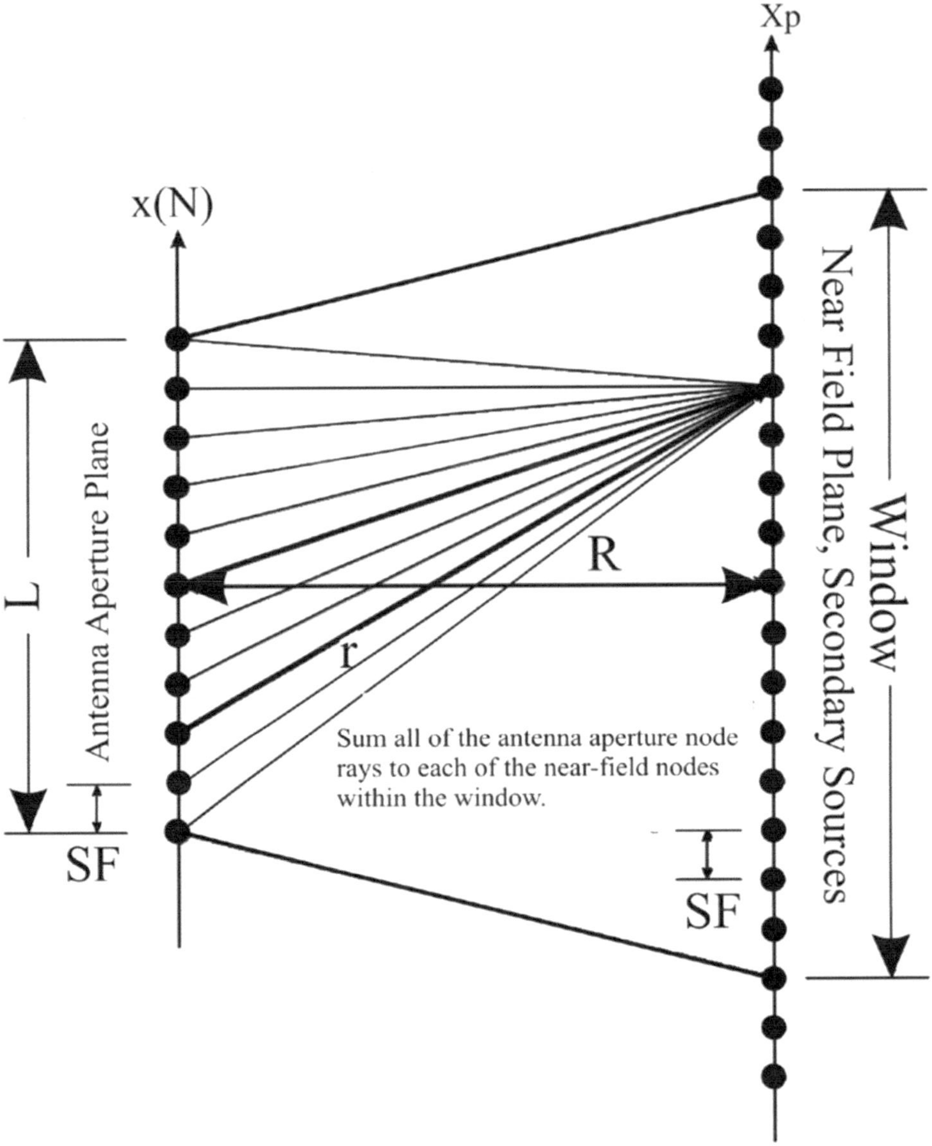

The first offset graphic shows the power distribution at the face of the antenna. Notice the amplitude rolls off toward the edges, but the phase is uniform.

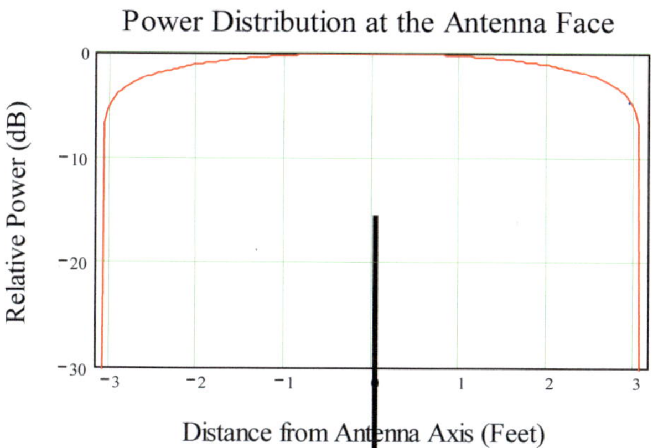

This is the power distribution over the face of the antenna for all of the following slides. Note that the power is down by about 5 dB at the edges of the 6 foot dish.

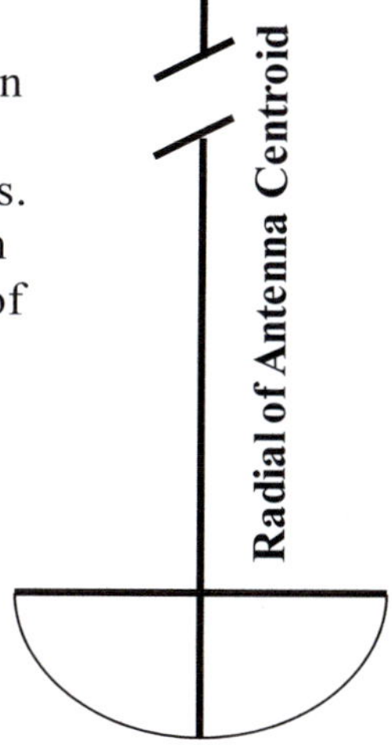

Antenna Diameter = 6 feet
Frequency = 11.8 GHz
Wavelength = 1.0 inches

This is the wave front a mere one foot in front of the antenna. Notice the variations in amplitude. There is similar variation of phase, but that is not plotted.

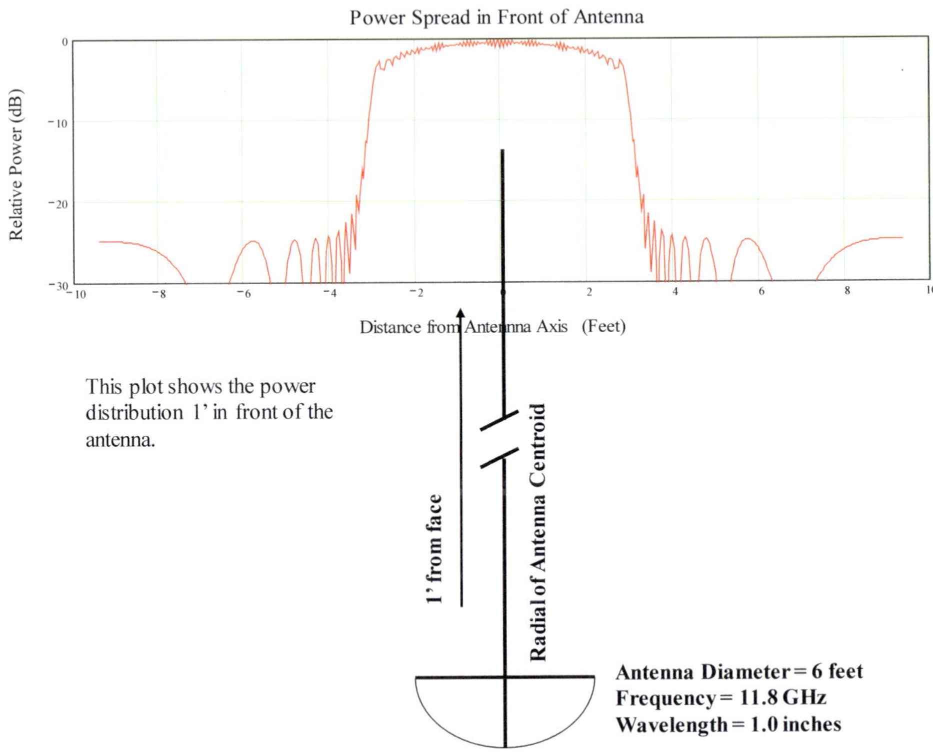

This plot shows the power distribution 1' in front of the antenna.

The wave front is now three feet from the face of the antenna.

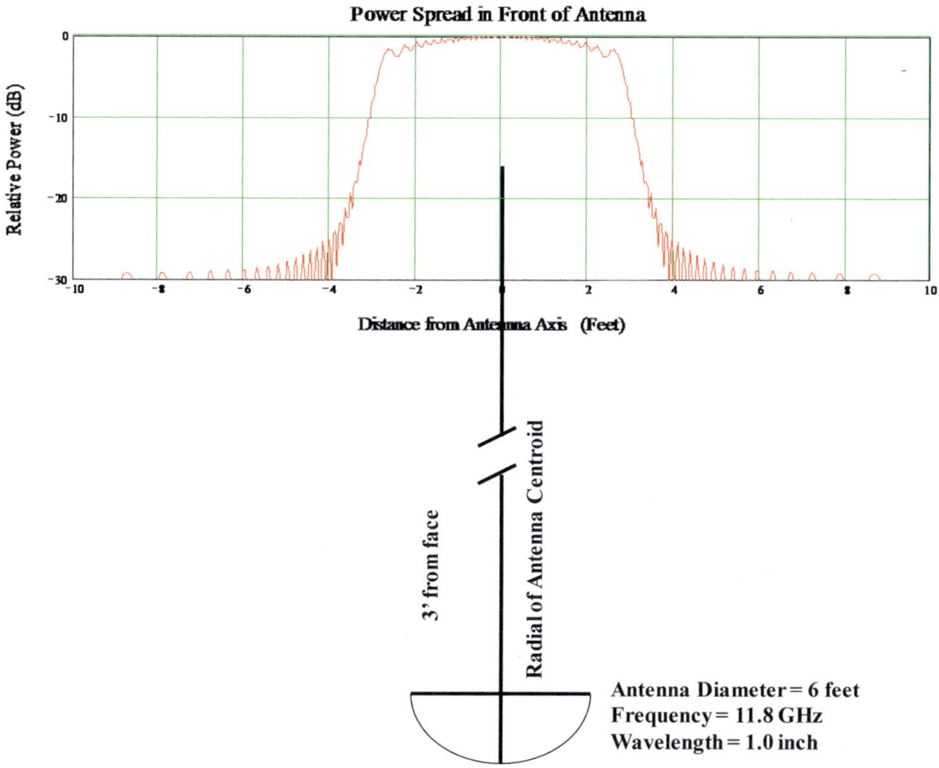

This has no resemblance to the shape of the far-field pattern.

The offset here is 12 feet..

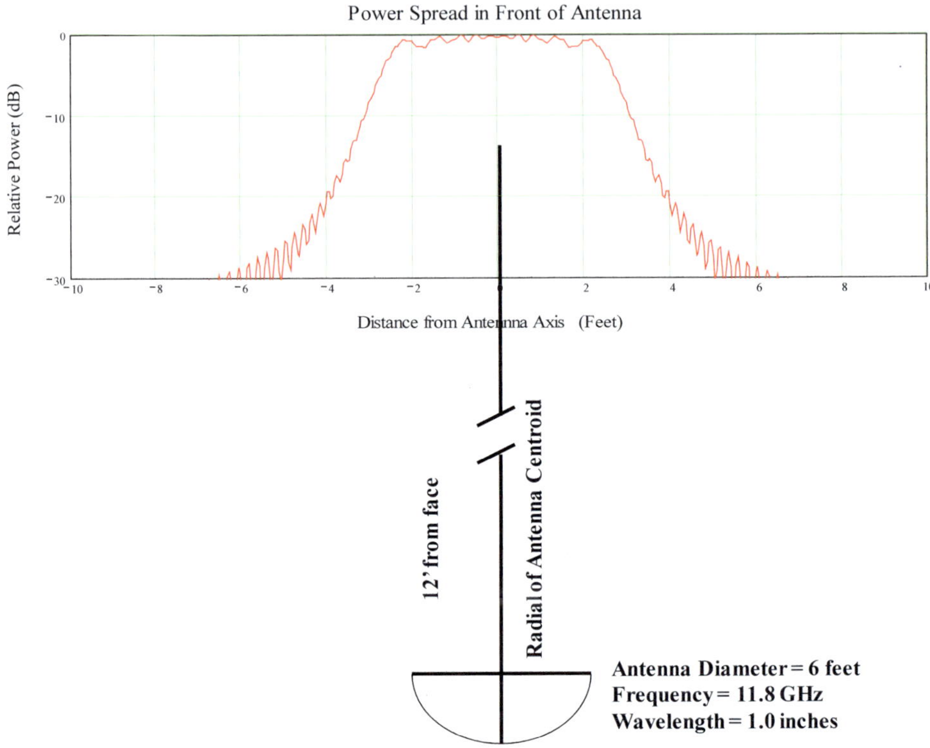

Here at 25 feet the shape is beginning to look like an antenna pattern. A receive antenna that this distance would receive considerable power.

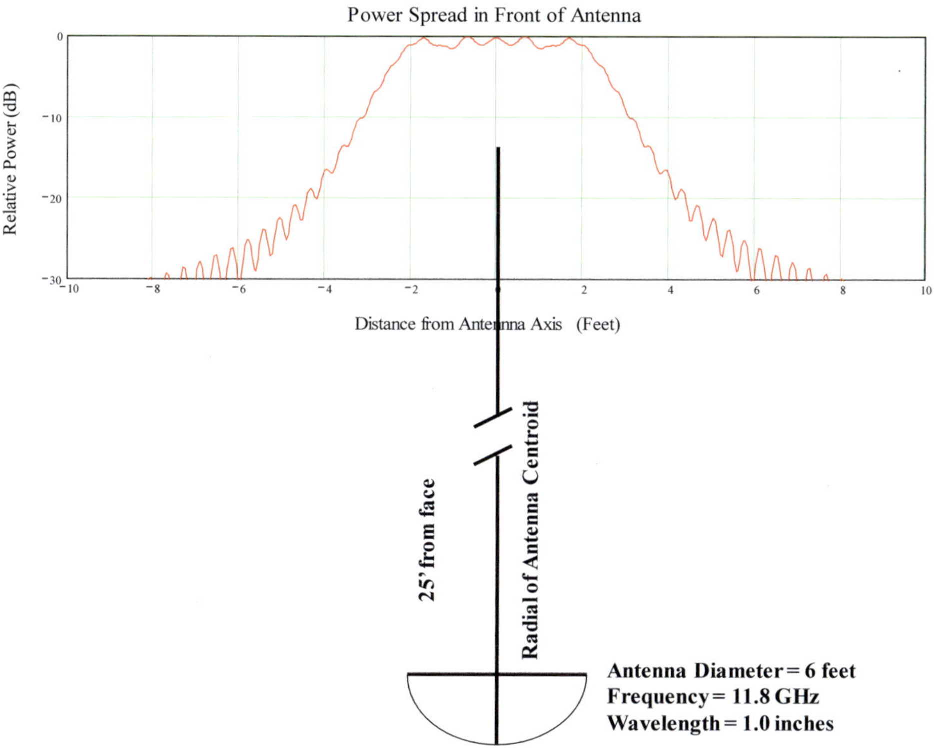

At 100 feet, the side lobes are beginning to form.

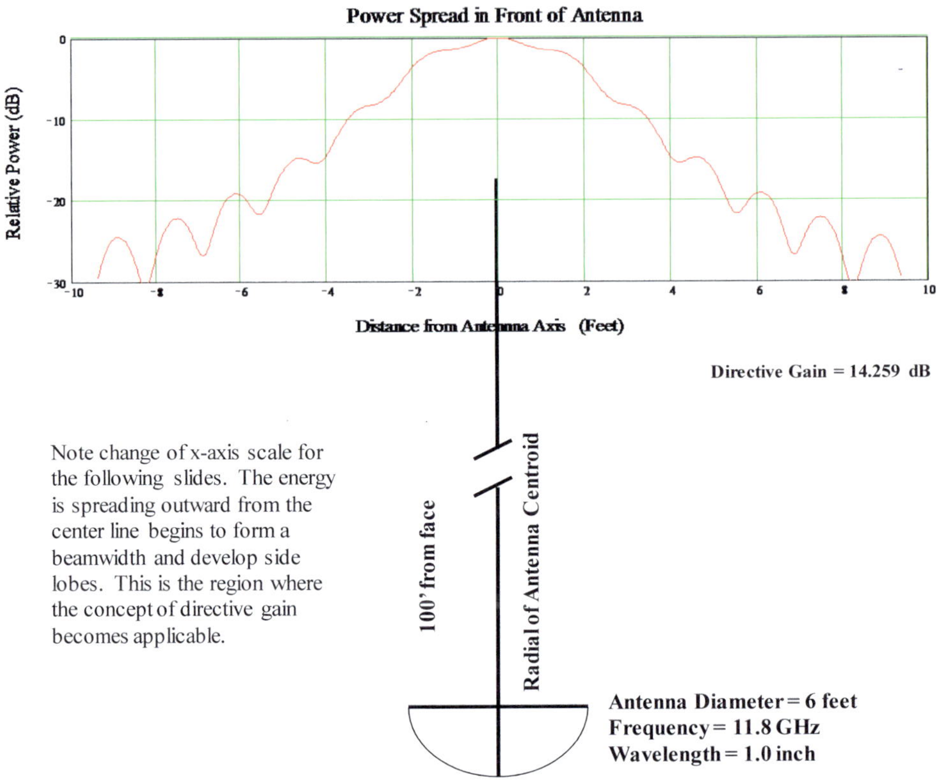

Note change of x-axis scale for the following slides. The energy is spreading outward from the center line begins to form a beamwidth and develop side lobes. This is the region where the concept of directive gain becomes applicable.

At 432 feet the side lobes are really forming up. But notice that the lateral distance between the firsts nulls is only 13 feet. This shows that the radiation in travelling in a very narrow space much like flowing down a pipe.

This is the case where the far-field is defined as D^2/λ. Sometimes designers may call this "good enough".

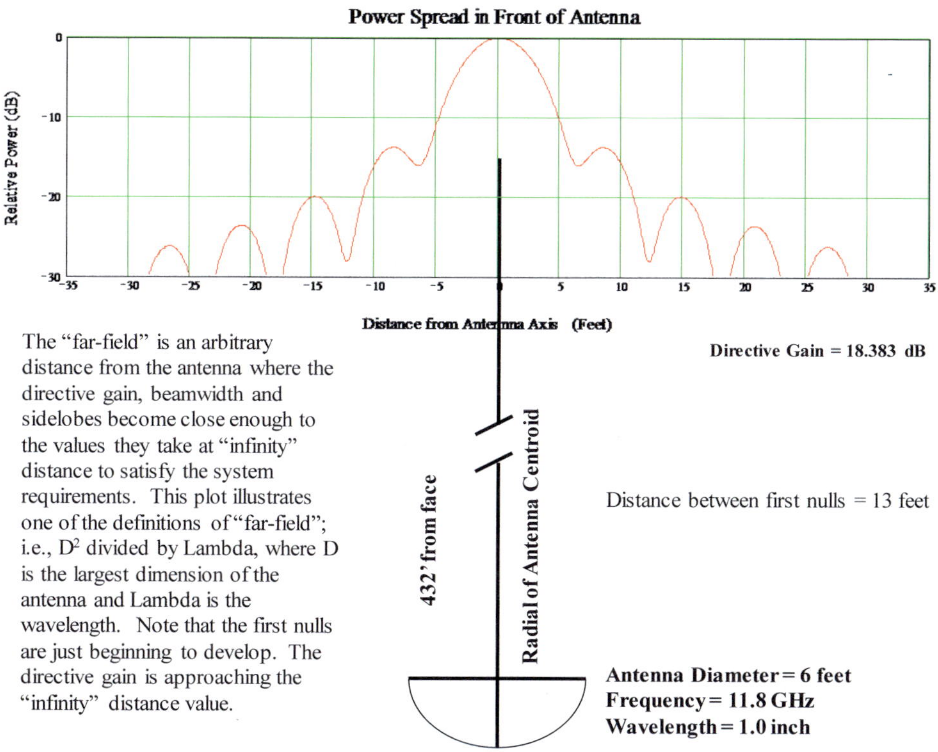

The "far-field" is an arbitrary distance from the antenna where the directive gain, beamwidth and sidelobes become close enough to the values they take at "infinity" distance to satisfy the system requirements. This plot illustrates one of the definitions of "far-field"; i.e., D^2 divided by Lambda, where D is the largest dimension of the antenna and Lambda is the wavelength. Note that the first nulls are just beginning to develop. The directive gain is approaching the "infinity" distance value.

Here is the conventional $2D^2/\lambda$ distance. Notice that the side lobes have not fully formed. Also notice difference between the 1D and 2D cases is minimal. Some designers get very nervous if they cannot get the 2D, no need. This why I tell people that the rule for "far-field" is not a hard limit.

The distance is about 0.16 miles, but the distance between nulls is only 26 feet, only four times greater than the diameter of the antenna!!

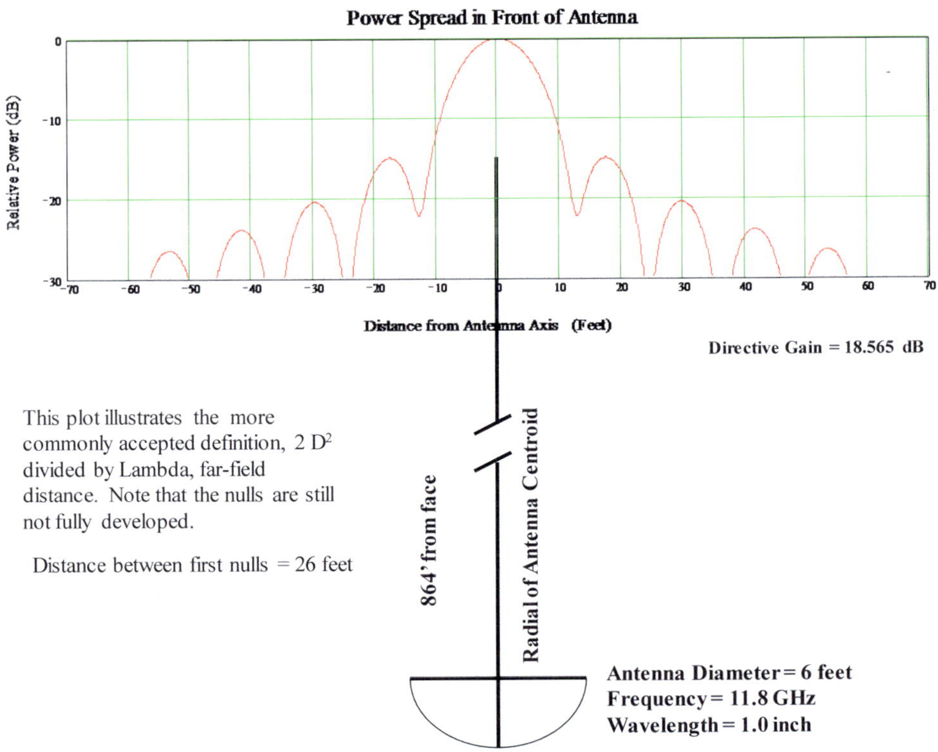

Now we are out to 3456 feet away. Look at the very deep nulls and the beamwidth is fractionally narrower.

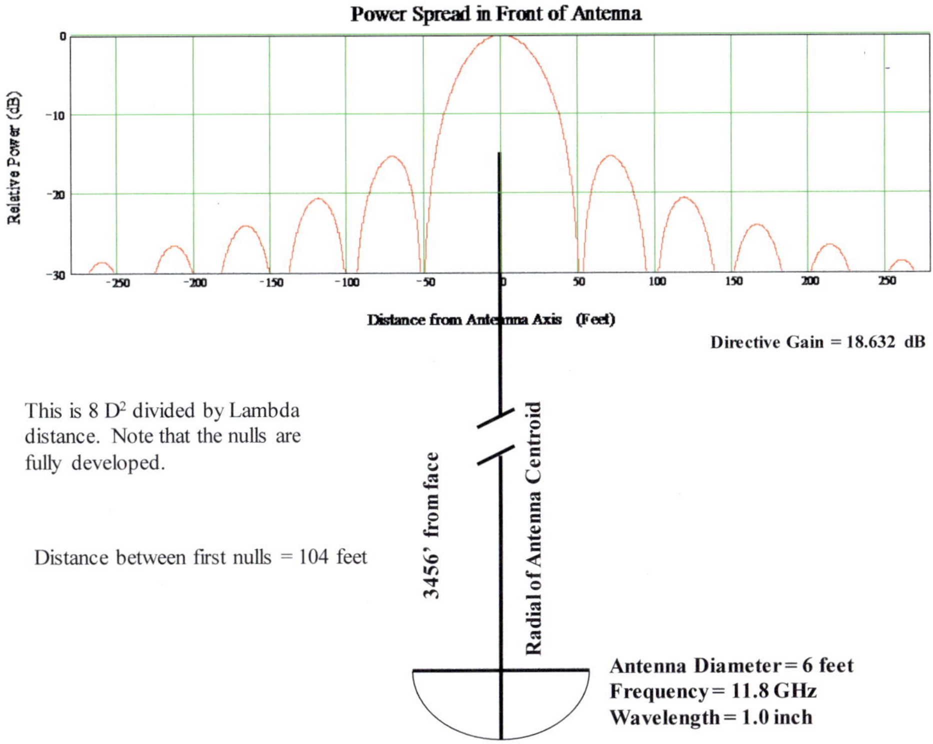

Finally, the radiation pattern in the very, very far-field. Notice the change of scale on the y-axis. The 3 dB beamwidth is about one degree so the antennas must be pointed very accurately.

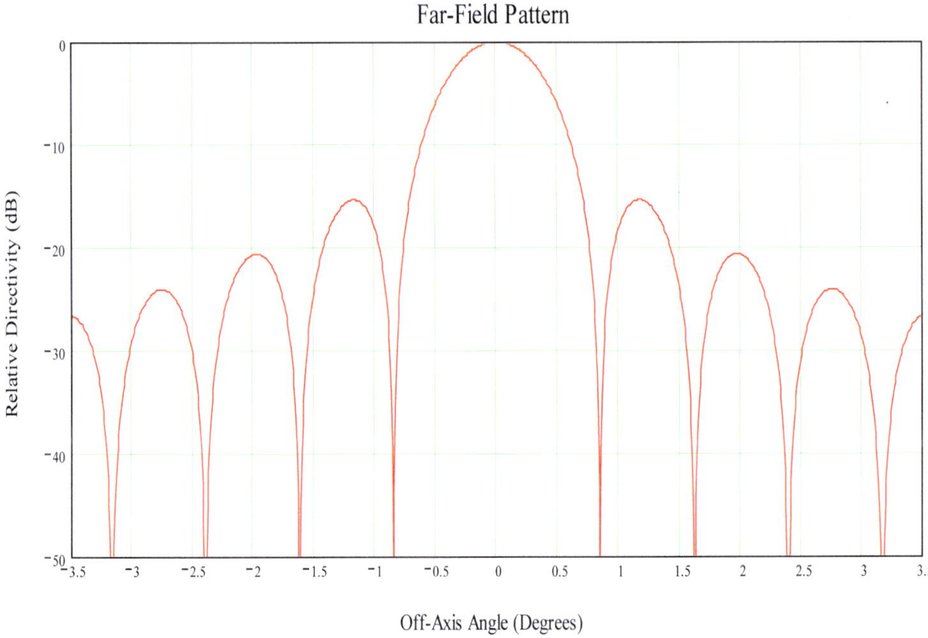

Here is the pattern in terms of angular offset from centerline.

Notice that the first side lobe is lower than -13.6 dB. That is because the amplitude distribution across the face is not uniform.

Propagation Link

So what is the mathematics for getting a known transmitted signal to the power level at the input of the receiver? Let's use this geometry to go through the procedure.

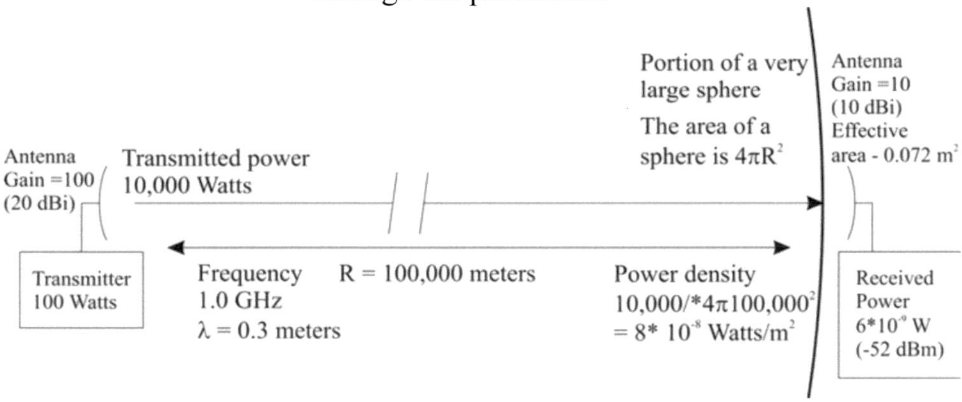

The level of the transmitted power is the transmitter power times the antenna gain (10,000 Watts). The power loss due the signal spreading out to the sphere is $1/4\pi R^2$. So, the power density at the receive antenna is $8{*}10^{-8}$ Watts/m^2. Notice that up to this point, frequency was not a factor. Now, the effective area of the receive antenna is: $Ae = Gr\lambda^2/4\pi = 0.072\ m^2$ which yields $6{*}10^{-9}$ Watts (-52 dBm)[43]. In the world of receivers, that is a very strong signal.

Here is very interesting thing I learned many decades ago. I do not remember if it was original with me or whether it was told to me.

Set up two omni antennas (D = 1.0) placed R distance apart.

The power density on the surface of the sphere is $1{*}1/(4\pi R^2)$ W/m^2. The area of the receiving omni is $1{*}\lambda^2/4\pi$. So, the received power is

[43] dBm is dB below one milliwatt. 100,000 meters is about 62 miles.

$\lambda^2/(R^2 * 4\pi * 4\pi)$ Watts. Then let $R = 1 * \lambda$. The power received, in dB, is $10 * \log(1/(4\pi)^2) = -22$ dB. Stated in words, the attenuation between two omni-directional antennas spaced one wavelength apart is 22 dB. We know that spreading in a function of $1/R^2$ or 6 dB each time the distance is doubled.

This is a very powerful tool. Let's take 1 GHz frequency as an example. The wavelength is approximately one foot. So:

1 foot -22 dB
2 feet -28 dB
4 feet -34 dB
8 feet -40 dB
16 feet -46 dB
32 feet -52 dB
64 feet -58 dB
128 feet -64 dB
256 feet -70 dB
512 feet -76 dB
1000 feet -82 dB
2000 feet -88 dB
4000 feet -94 dB
8000 feet -100 dB
3 miles -106 dB
6 miles -112 dB
12 miles -118 dB
24 miles -124 dB
48 miles -130 dB
96 miles -136 dB
200 miles -142 dB
400 miles -148 dB
800 miles -154 dB
1600 miles -160 dB
3200 miles -166 dB
6400 miles -172 dB
12800 miles -178 dB
25600 miles -184 dB
51200 miles -190 dB
102400 miles -196 dB
204800 miles -202 dB (The moon)

You can do this very quickly either mentally or on a scratch pad. Then just add the transmit and the receive antenna gains, in dB, and, voila, you have a very good approximation of the propagation link!!!

Said in words**: 22 dB for the first wavelength and 6 more dB for each doubling**. Caution: In actual practice, if you put two antennas only one wavelength apart you probably would not be in the far-field and this equation would fail. It is intended to estimate propagation attenuation to the far-field.

Skin Depth

Direct current (dc) flows uniformly throughout the volume of a wire. But an alternating current creates fields inside the wire that pushes the current to the outer edge. At very high frequencies, the electron flow (current) is confined to very near the edge of the wire.

This effect occurs in any conductor, not just wires. That is why you can plate the inside of waveguide with silver to improve performance over aluminum. Virtually all of the current is in the thin layer of silver.

The EM name for this property is "skin depth". At one "skin depth" the value of the voltage is 36.8% of the level at the surface. That is -8.2 dB. The equation for the depth that the current extends into the metal is:

Permeability of a Good Non-Magnetic Conductor

$$\mu_0 = 1.257 \frac{\mu H}{m}$$

Conductivity of Copper

$$\sigma \equiv 58 \cdot 10^6 \cdot \frac{mho}{m}$$

One Skin Depth

$$\delta(f) \equiv \frac{1}{\sqrt{f \cdot \pi \cdot \mu_0 \cdot \sigma}}$$

$$\delta(60 \cdot Hz) = 0.3\,in$$
$$\delta(1 \cdot MHz) = 2.6 \times 10^{-3}\,in$$
$$\delta(1 \cdot GHz) = 8.2 \times 10^{-5}\,in$$
$$\delta(10 \cdot GHz) = 2.6 \times 10^{-5}\,in$$

So, clearly, at high frequency plating does not have to be very thick to contain virtually all of the current.

Some people recommend a plating thickness of three skin depths to have minimal conduction loss. Other people recommend five skin depths.

Antenna Arrays

An antenna array is simply a "large" antenna composed of two or more smaller antennas, called "elements". The array may be one dimension such a line of element. It may be two dimensional as elements arranged on a flat rectangle, or it may be three dimensional as elements placed in a spherical shape.

The elements are antennas in their own right and maybe monopoles, dipoles, horns or any other type.

The most important parameter in array design is the spacing between the elements. A common spacing is ½ wavelengths, but other spacing are also used. When two elements are spaced far apart, it is called an interferometer.

Recall this from an earlier section.

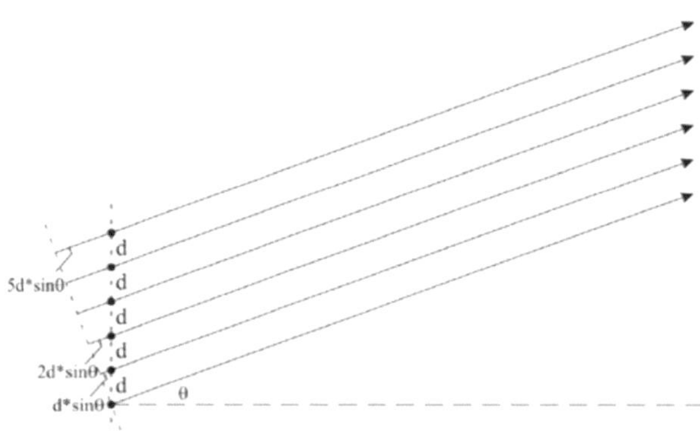

Let $d = \lambda/2$

$$V_i \equiv \left[\left[\frac{\sum_{n=0}^{5}\left(e^{j\cdot 2\cdot \pi\cdot n\cdot d\cdot \sin(\theta_i)}\right)}{6}\right]\right]$$

$$P \equiv \overrightarrow{\left(10\cdot \log\left(V\cdot \overline{V}\right)\right)}$$

It has been modified in two ways. The summation begins at the end instead of the center.[44] The elements are now individuals instead of part of a

---

[44] Summing from the end introduces a phase shift in the far-field. Generally, we do not care about the phase.

solid antenna. In this case, each element is an omni directional antenna. The pattern we get is called the array factor and the directivity is the "array directivity". In real life the elements must be some type of antenna; say a dipole. More on that later, for now we are going to concentrate on the array pattern, which for this case is:

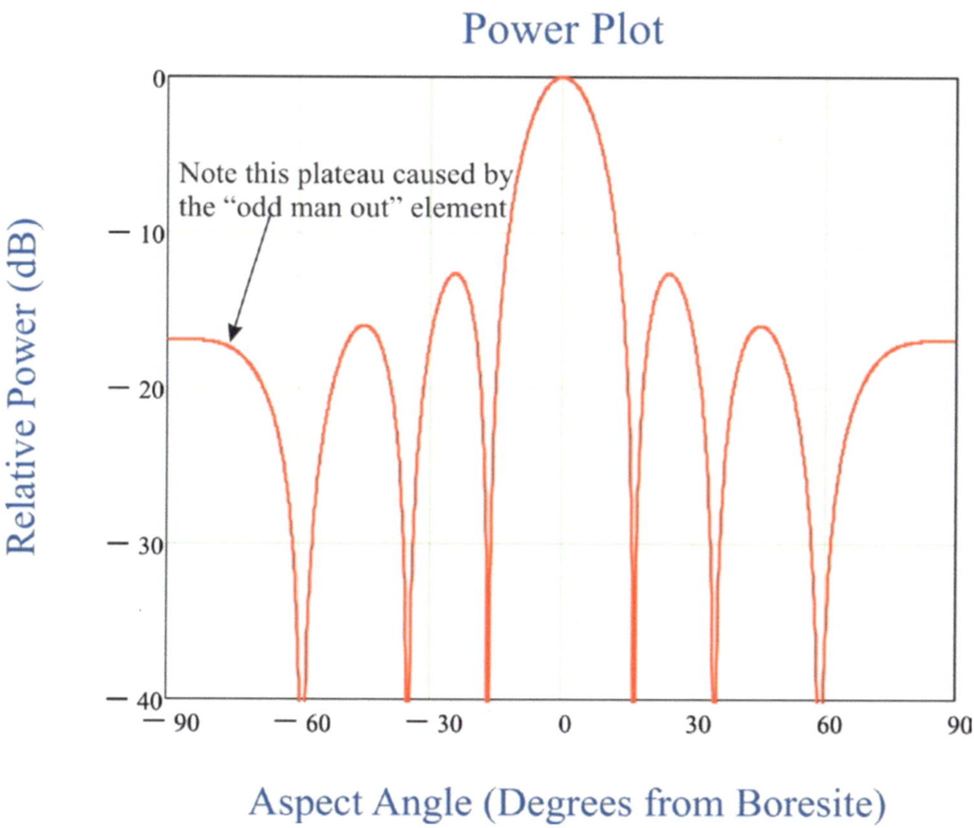

Notice the plateau at 90 degrees angle. That is caused by have an odd number of elements. On each side of center, element pairs enforce each other, leaving the one unpaired one in the middle. Now reduce the number of elements to six, no plateau.

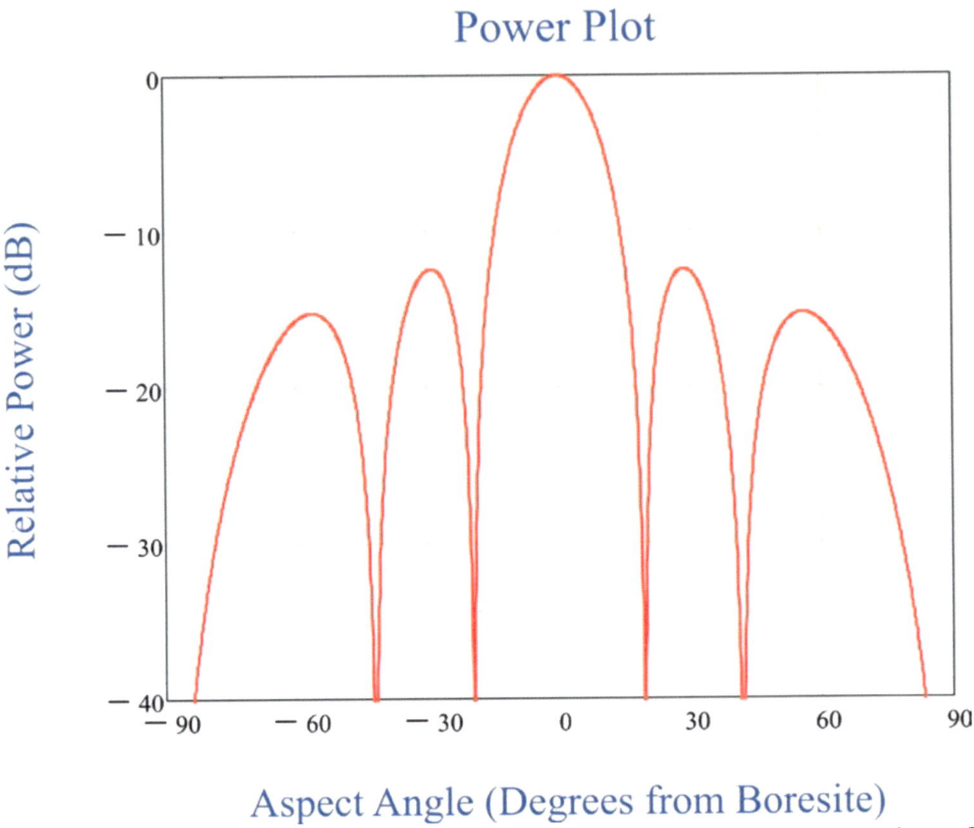

There is much to be learned from the operation of a two element array.

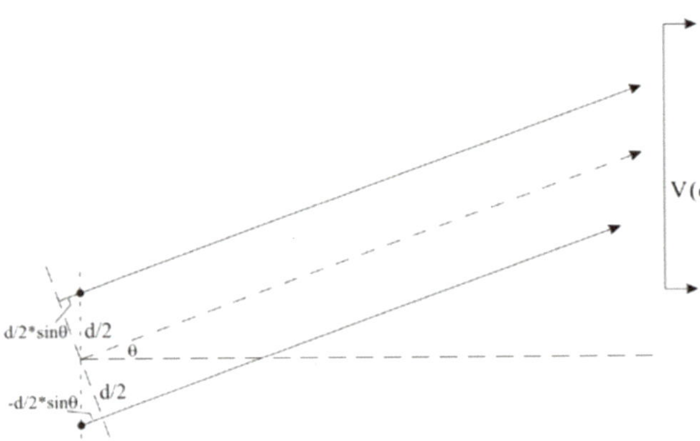

$$V(d) \equiv \frac{e^{-j\cdot 2\cdot \pi\cdot \frac{d}{2}\cdot \sin(\theta)} + e^{j\cdot 2\cdot \pi\cdot \frac{d}{2}\cdot \sin(\theta)}}{2}$$

$$P \equiv \overrightarrow{\left(10\cdot \log\left(V\cdot \overline{V}\right)\right)}$$

Here is a plot for various values of d, in wavelengths.

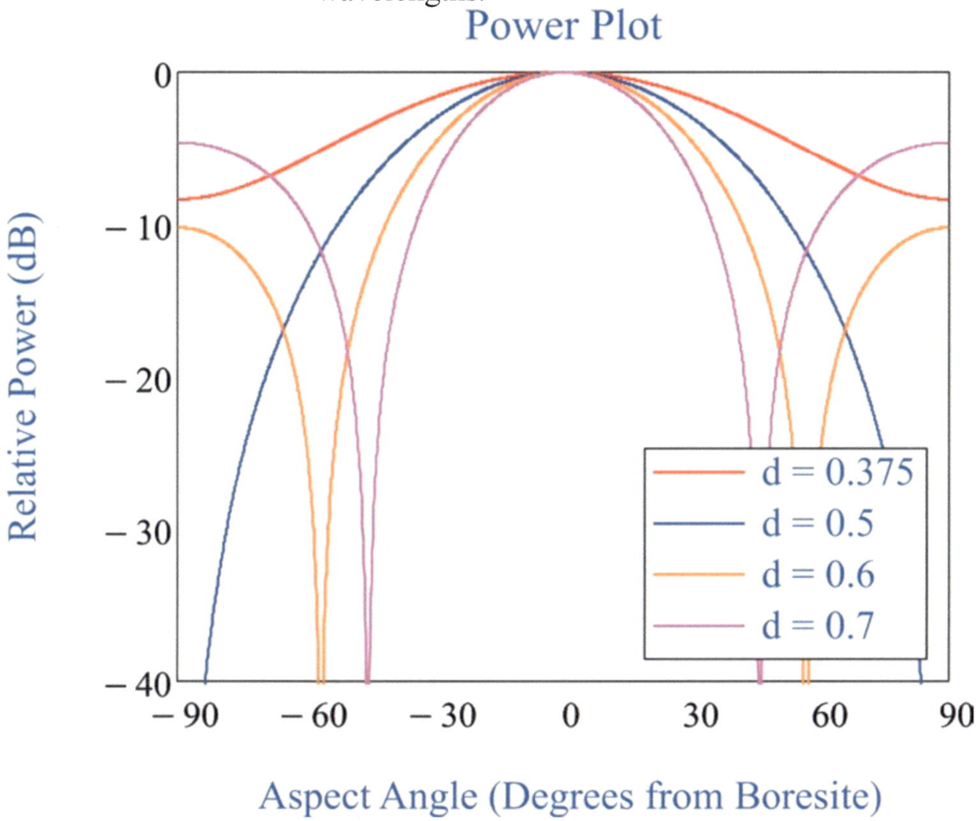

Notice that the sum is 1.0 in every case. But, we learned earlier that narrower beamwidth means higher gain. This means that the **value of the summation tells you nothing about the gain of the array.** As hard as it is to believe, the different beamwidths result from mutual coupling effects.

So, how do we find the gain of the array? The only precise way is to integrate the power over sphere surrounding the array. Or we can use the approximation formula discussed in the section on directivity.

By using my directivity equation we get:

$d = 0.375\lambda$

$$10 \cdot \log\left(\frac{360}{360 \cdot \sin(41.9 \cdot deg)}\right) = 1.8 \;\; dBi$$

$d = 0.5\lambda$

$$10 \cdot \log\left(\frac{360}{360 \cdot \sin(30 \cdot deg)}\right) = 3 \;\; dBi$$

$d = 0.6\lambda$

$$10 \cdot \log\left(\frac{360}{360 \cdot \sin(24.6 \cdot deg)}\right) = 3.8 \;\; dBi$$

$d = 0.7\lambda$

$$10 \cdot \log\left(\frac{360}{360 \cdot \sin(20.9 \cdot deg)}\right) = 4.5 \;\; dBi$$

Now, let's take the case where d = 1λ.

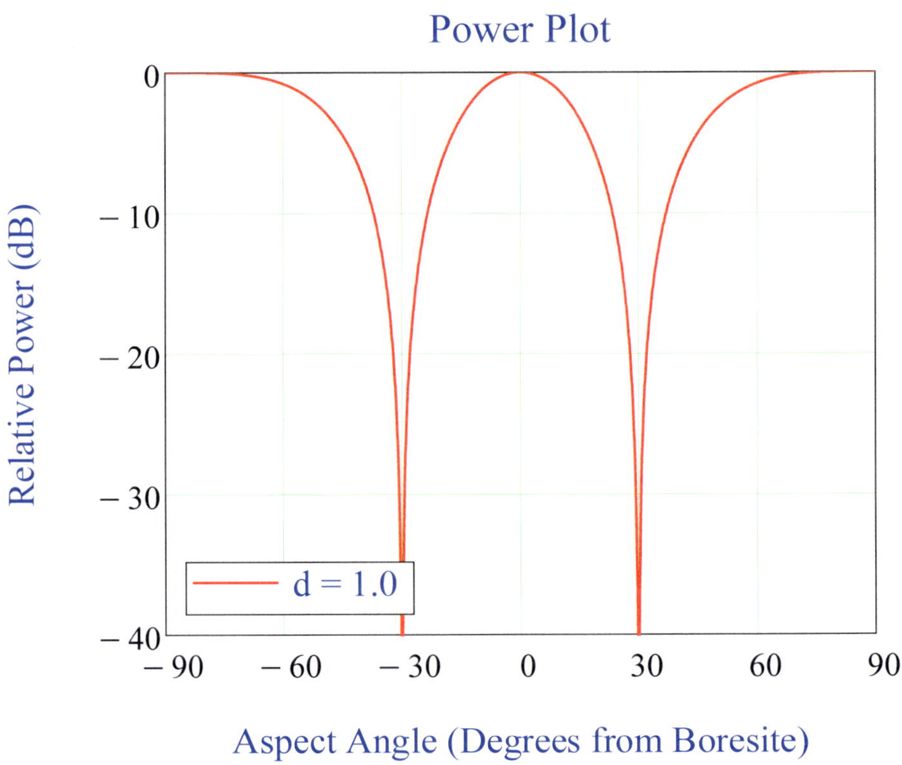

Now the pattern has a peak at ±90 degrees. This peak is called the grating lobe. Remember that this pattern results from omni-directional elements. If the real element does not have gain in that direction, such as a dipole, the grating lobe will be suppressed.

The next step is to separate the elements by several wavelengths.

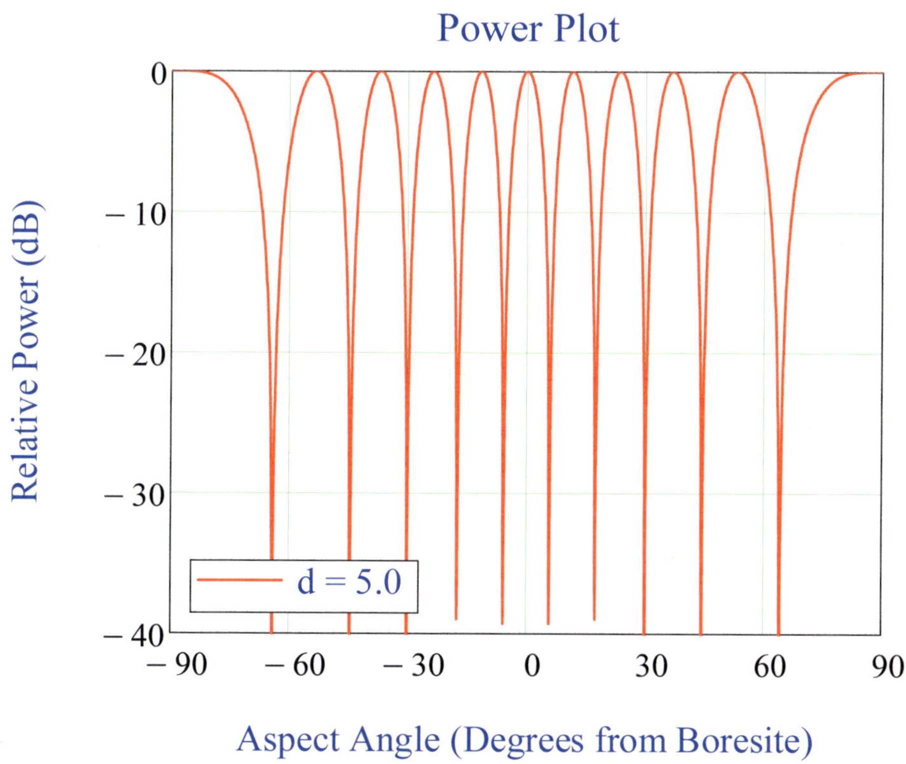

This is the pattern of an interferometer. There are five peaks each side of center.

At this point I want to caution about a very common mistake. The elements of an array have their own directivity. You will find claims (even in text books) that the total directivity is the element directivity multiplied by the array directivity. **That is not true!** Here is an excerpt from my paper on directivity.

Directivity and Pattern Multiplication

A common method of estimating array gain is to multiply the element gain by the array factor.

This implies: $D_T = D_a \times D_e$

and,

$$D_T = \frac{4\pi}{\iint U_a(\theta,\phi)\sin\theta \, d\theta d\phi} \times \frac{4\pi}{\iint U_e(\theta,\phi)\sin\theta \, d\theta d\phi}$$

However, this equation is not true. In some cases, it is not even close. Actual combined directivity may be more or less than the simple product.

The correct expression is:

$$D_T = \frac{4\pi}{\iint U_a(\theta,\phi) U_e(\theta,\phi)\sin\theta \, d\theta d\phi}$$

Where $Ua(\theta,\phi)$ is the pattern of the array and $Ue(\theta,\phi)$ is the pattern of the element.

All of the examples in this section have used a linear array. The extension to two and three dimensional arrays is straightforward and can be found in many text books. My intent here is to lay the ground work and present content you may not find in text books.

Here some more excerpts from my paper on directivity.

My Paper on Directivity

Directivity and the Summation Equation

Summation in the Far Field

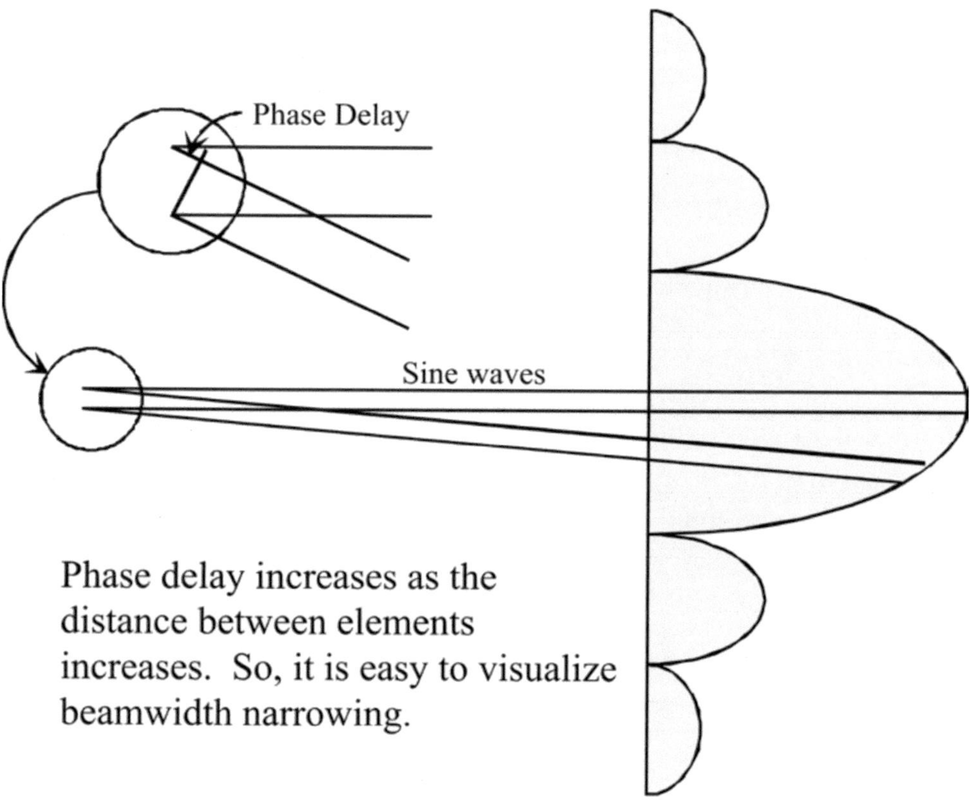

Phase delay increases as the distance between elements increases. So, it is easy to visualize beamwidth narrowing.

However, it is not so obvious why the radiation intensity changes!!!

Directivity of Discrete Arrays

$$F(\theta,\phi) \triangleq \sum_{n=1}^{N} A_n e^{jk(x_n \sin\theta\cos\phi \; + \; y_n \sin\theta\sin\phi)} \quad \text{(Voltage)}$$

$$U(\theta,\phi) \triangleq [F(\theta,\phi)]^2 \quad \text{(Power)}$$

$$D(\theta,\phi) \approx \frac{4\pi}{\sum\sum U(\theta,\phi)\sin\theta\, d\theta d\phi}$$

At the peak-of-beam

$$F(\theta,\phi) \triangleq \sum_{n=1}^{N} A_n e^{j0} = \sum_{n=1}^{N} A_n \overset{A=1}{=} N$$

$$U(\theta,\phi) \triangleq [F(\theta,\phi)]^2 = \left[\sum_{n=1}^{N} A_n\right]^2 \overset{A=1}{=} N^2$$

Array Improvement
(Over one element of the array)
(Power into each element is reduced by 1/N)

$$\triangleq \frac{\left[\sum_{n=1}^{N} A_n\right]^2}{N} \overset{A=1}{=} N$$

So, the directive "gain" of an array is N times the value of the element directivity (in the array); assuming all elements have the same value. But there is a "gotcha", see the next slide.

Note: As I first began the study arrays, I tortured myself whether to divide the sum by N or N^2. It was when I wrote my paper on directivity that I realized that the answer is “neither”. See the second equation on the next page. You can normalize the value in the far-field to any number you like. I like 1.0.

Effect of Element Spacing

Consider the summation equation

$$F(\theta,\phi) \triangleq \sum_{n=1}^{N} A_n \, e^{jk(x_n \sin\theta \cos\phi \; + \; y_n \sin\theta \sin\phi)}$$

At peak-of-beam this equation yields the same value for any element spacing.

However, we know that element spacing affects directivity; and therefore, radiation intensity at peak-of-beam.

So, the summation equation should be expressed as:

$$F(\theta,\phi) \triangleq \sum_{n=1}^{N} A_n(d) e^{jk(x_n \sin\theta \cos\phi \; + \; y_n \sin\theta \sin\phi)}$$

A(d) changes with element spacing because it is a function of mutual coupling!!!

As stated the array "gain" is N times A(d), but total gain is a function of d.

A(d) of a single element is the same even if other elements are match terminated instead of actively driven.

Directivity vs Element Spacing

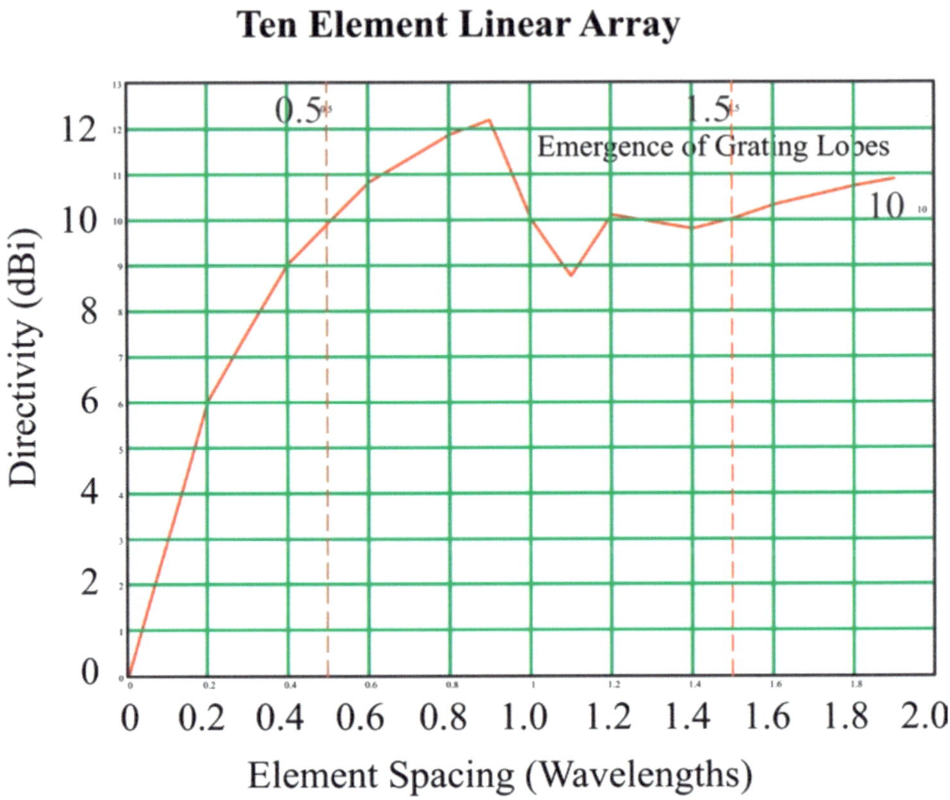

The lower element spacing is impossible to achieve physically; and, if achieved, have high losses due to high currents of mutual coupling.

Polarization

After you study this section, you will know more about polarization than 90+ percent of your peers. There are two reasons.

First, very little attention is given to polarization in our academic programs.

Second, a vast majority of antenna work involves transmit and receive antennas having the same polarization, so the subject is moot.

If the transmit and receive antennas do not have the same polarization "polarization mismatch loss" occurs.

As I start this section, I feel compelled to remind you that an EM signal is a propagating sine wave and, although sine wave signals are not vectors, they can be treated as if they are. So, when I draw a line with an arrow point, it represents a sine wave oscillating at the signal frequency. The length of the pointed line is the magnitude in volts. The voltage may be expressed in peak value or rms value or any other value. Your choice. The angle of the "vector" to some reference "vector" is the phase of the sine wave relative the phase at the reference, usually zero. I emphasize this because it is easy to look at the "vector" and forget about the signal frequency.

Also, remember that although EM waves have both a magnetic and electric component, we use only the electric component (voltage) in our analysis. The orientation in space of the E-vector is the same direction as the current that created it.

The analysis of polarization almost always is done by resolving the E-vector into two orthogonal components such the vertical and horizontal. But remember **<u>there is only one E-vector</u>** and we decompose it for ease of measurements and computations.

There are two cases where polarization mismatch is very critical. One is a non-stationary three dimensional array.[45] (In the far-field an array is regarded as a single point antenna.) The other is distortion of the polarization of a signal as it passes through a radome (next section).

Polarization Types

Let's define polarization. In this figure a single E-vector is travelling to the right.[46] It maintains the same position in space as it travels. It is a real thing. Just because you cannot see or feel it does not mean it does not exist. It is energy on the move. If you stand in front of it, it part of it will pass through your body and part will be dissipated as heat in your body. If it is intense enough, it will burn you, just like a piece of beef in the microwave oven. This is linear polarization.[47]

---

[45] As an example, I have placed antennas on the circumference of a cylindrical missile. As the missile moves and rotates, the relative positons of the antennas change causing the composite polarization to change.

[46] The illustration depicts one vector as it moves along the line. It is not a bunch of vectors!

[47] This vector not necessarily vertical. It can point in any direction, but it maintains that same direction as it travels

Linear Polarization Propagation

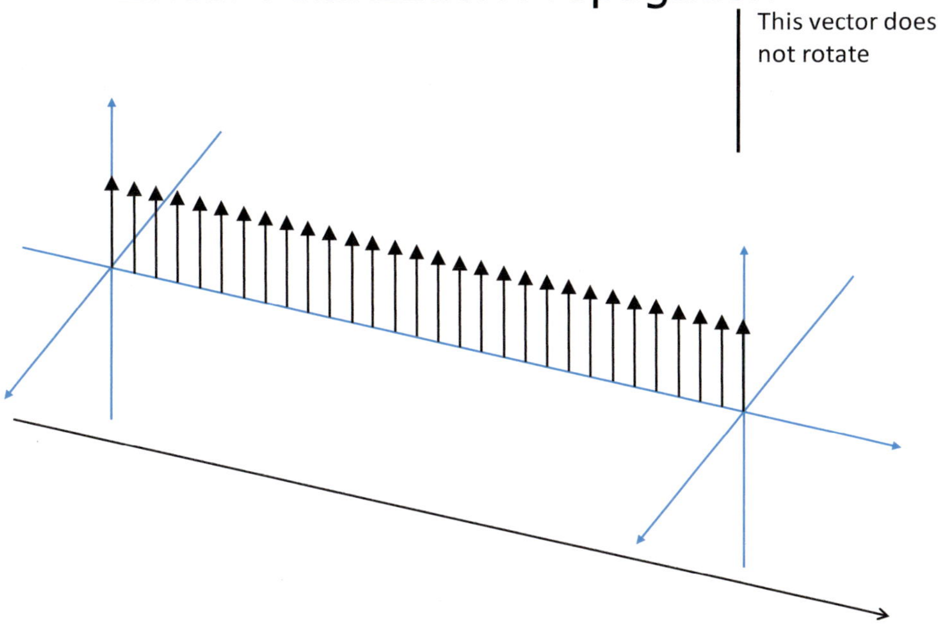

Now, let this vector not only comes toward you but let it rotate in space as it travels. (Like a cheerleader's baton twirling and coming toward you at 186,000 miles/second. Catch it if you can.) It rotates a full 360 degrees in one wavelength of travel. This circular polarization and it is right hand circular polarization (RHCP) if it is rotating ccw and left hand circular polarization (LLCP) if it is rotating cw as it comes toward you.[48]

I have made the special notation that there is only one polarization because people get used to the terms "vertical polarization" and "horizontal polarization", like they are two separate things. They are not. They are components of the **one** actual polarization.

[48] Go to Wikipedia for some animated graphics.

Circular Polarization Propagation

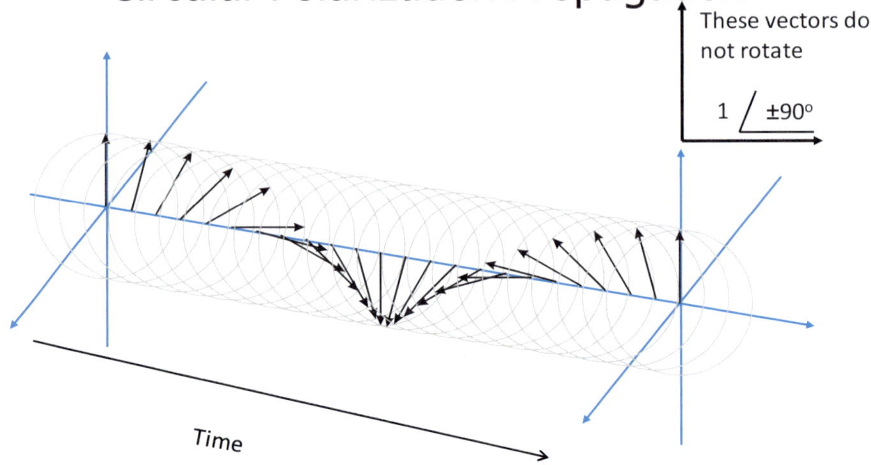

There is only one polarization

Finally, there is elliptical polarization. It is like circular except that the length of the cheerleader's baton is also changing as it travels. Again completing one full revolution in one wave length.

As you have probably deduced by now, linear and circular polarizations are special, limiting, cases of elliptical polarization.

Now, I am going to tell you something that seems pretty incredible. These rotating vectors can be represented by two orthogonal vectors that are not rotating. Look in the upper right hand corner of each figure for the stationary equivalent. For circular, it is two orthogonal vectors of the same length but 90 degrees time phase apart[49]. If either

[49] +90 degrees for RHCP and -90 degrees for LHCP.

the condition of equal amplitude and/or 90 degrees time phase is not met, the polarization is elliptical.

Elliptical Polarization Propagation

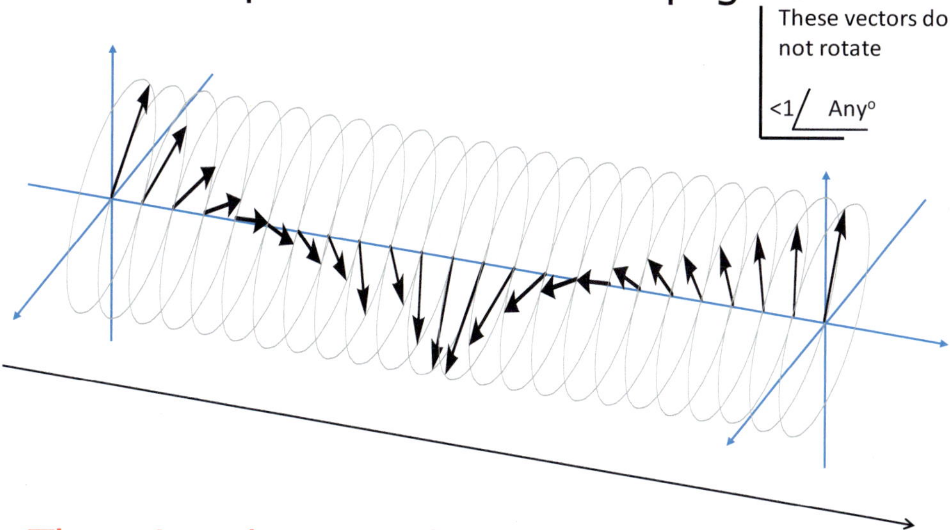

There is only one polarization

Polarization Ellipse

Most text books barely pay lip service to polarization. Almost all of them use this to describe polarization.

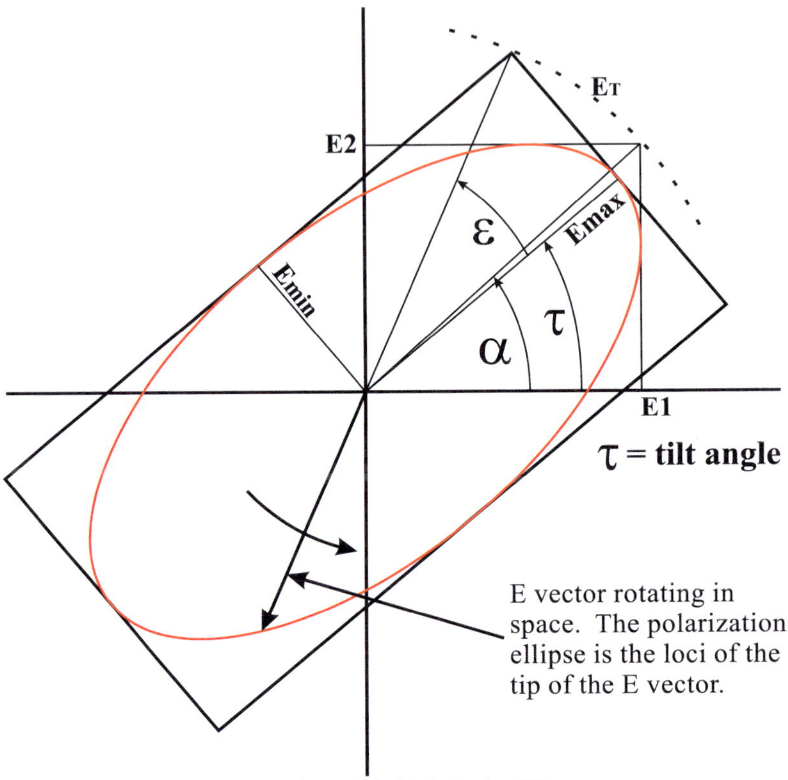

Equations of the Polarization Ellipse

$\alpha = \tan^{-1}(E_2/E_1)$; $\delta = \phi_{E2} - \phi_{E1}$

$\cos 2\alpha = \cos 2\varepsilon \cos 2\tau$

$\tan \delta = \tan 2\varepsilon / \sin 2\tau$

$\tan 2\tau = \tan 2\alpha \cos \delta$

$\sin 2\varepsilon = \sin 2\alpha \sin \delta$

The problem here is that the polarization ellipse is not a real thing. It represents an ac voltage of a propagation wave, but there is no physical way to measure it. You need an antenna to measure the value of power (voltage) in the wave. **From this graphic how do you determine E_2 and E_1,** from which both α and δ are derived? This is as far as most text book go!!

The polarization ellipse is almost useless. First, you need to know the value of E1 and E2 and most texts do not show you how to get it. Secondly, the trigonometric equations are complicated and it is difficult (impossible?) to resolve which quadrant you are working in.

In a practical sense, you need an antenna to measure the response of an antenna under test. By far the most common antenna used for this purpose is the linearly polarized horn that is rotated on its axis.

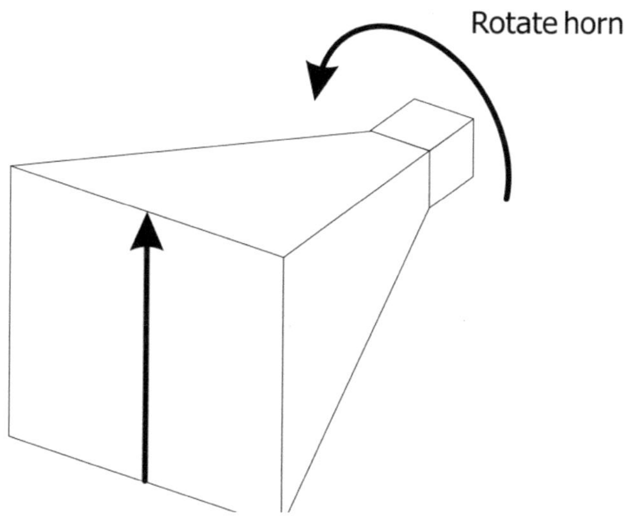

Set up a linearly polarized horn to measure the response from the antenna under test. Rotate the horn one full 360 circle in small increments. Convert the measured power to voltage. This is called the **"Polarization Pattern"**.

The polarization pattern

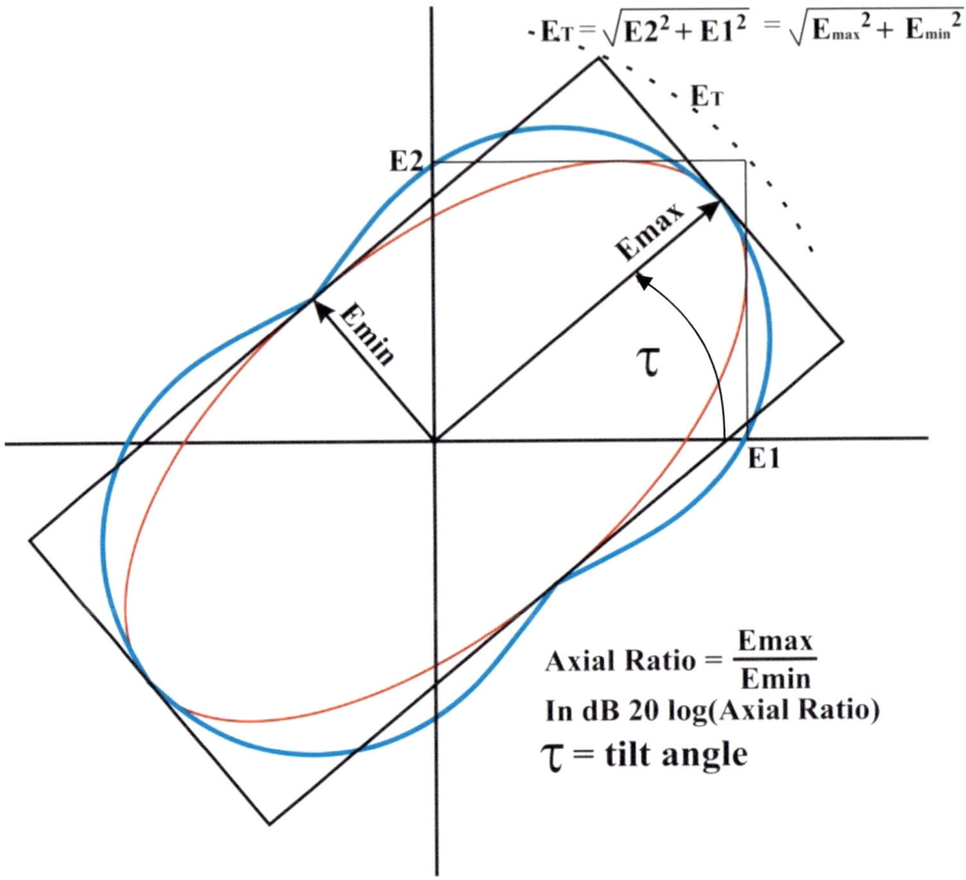

The Polarization Pattern

Plot the output of a linearly polarized antenna as it rotates on its axis to get maximum and minimum values. Since most antenna ranges measure power in dB, the axial ratio, in dB, is the difference between the peak and the valley of the plot.

Equations of the Polarization Ellipse

$$\alpha = \tan^{-1}(E_2/E_1);\ \delta = \phi_{E2} - \phi_{E1}$$

$$\cos 2\alpha = \cos 2\varepsilon \cos 2\tau$$

$$\tan \delta = \tan 2\varepsilon / \sin 2\tau$$

$$\tan 2\tau = \tan 2\alpha \cos \delta$$

$$\sin 2\varepsilon = \sin 2\alpha \sin \delta$$

Notice that when the rotation angle is on the horizontal and vertical axes, it yields the value of E1 and E2. Emax and Emin is the same for both cases. At least we can now solve the equations!! But only a few text books show the polarization pattern.

Why are we going to all of this trouble? We want determine how the polarization of the antenna under test affects our system. Assuming that the receiving antenna is not the same polarization, how does the polarization mismatch affect us? Not even the polarization pattern helps us very much in that regard.

But fear not, in 1892 Henri Poncair' invented his sphere, a graphical way to solve those tedious equations with absolute certainty of which quadrant you are in.

Look at this nomogram as a graphical way to solve an equation.

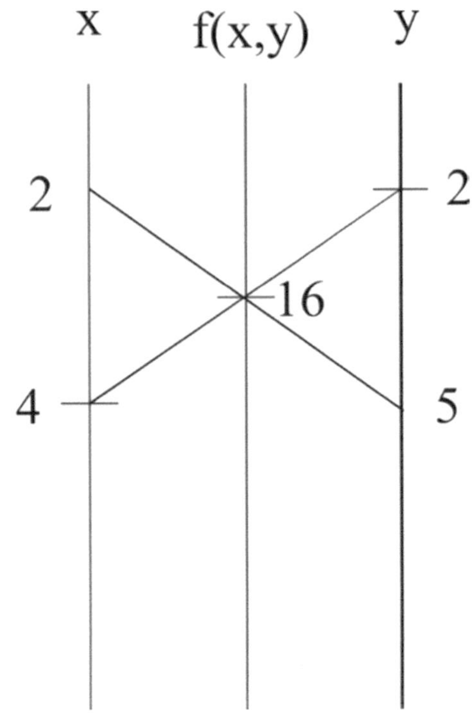

$$f(x,y) = 3x + 2y$$

Before the days of computers, it was quite common to solve equations with nomograms. An example would be to determine propagation loss based on frequency and distance.

The Poincare' Sphere

When people first see the Poincare sphere, they envision a sphere containing an antenna pattern. Disavow them of that notion at once. The Poincare' sphere is precisely like a nomogram; i.e., a graphical way to solve equations!

Poincare' Sphere

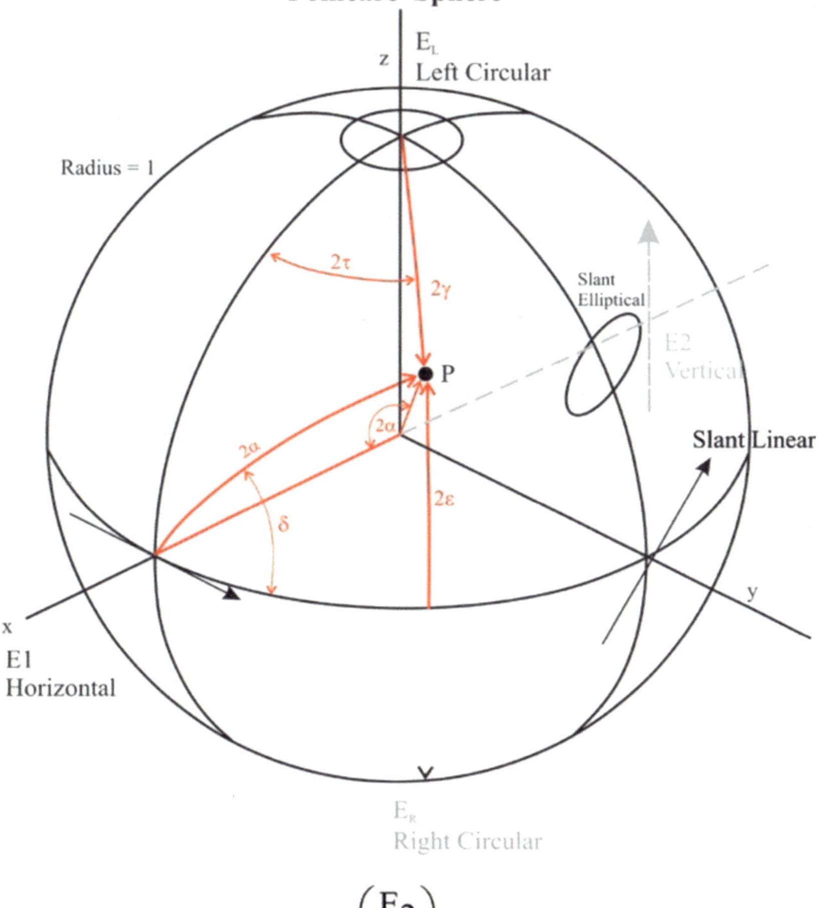

$$\alpha = \operatorname{atan}\left(\frac{E_2}{E_1}\right) \qquad \delta = \phi_{E2} - \phi_{E1}$$

$$\cos(2\alpha) = \cos(2\varepsilon)\cdot\cos(2\tau)$$
$$\tan(\delta) = \tan(2\varepsilon) \div \sin(2\tau)$$
$$\tan(2\tau) = \tan(2\alpha)\cdot\cos(\delta)$$
$$\sin(2\varepsilon) = \sin(2\alpha)\cdot\sin(\delta)$$

We discussed previously linear, circular and elliptical polarization. The Poincare' sphere displays every possible polarization. Linear polarization is on the equator. LHCP is at the North Pole and RHCP is at the South Pole. Elliptical polarization is everywhere else. Now, we can solve the equations. Using 2α and δ places the polarization point (P) on the sphere. By spherical trigonometry, you can find all of the other values.

What is the primary thing we want to know? We want to know how much signal power is lost because the polarizations of the transmitter and receiver antennas do not match!![50] The Poincare' sphere has this wonderful property; i.e., if the angle between the two polarizations is 2θ, then the polarization mismatch loss is $\cos(\theta)^2$!![51] Expressed in dB it is $10*\log(\cos(\theta)^2)$. Take the case when the two polarizations are directly opposite on the sphere. Then 2θ is 180 degrees and θ is 90 degrees. The loss is infinity. This is the cross polarized condition and illustrates that **when the receive antenna is cross polarized with the transmit antenna, no power is received!** Furthermore, every possible

---

[50] Remember, that in the far-field, an array of antennas appear as a single antenna with its corresponding polarization.

[51] It is very common to hear people say the mismatch between linear and circular polarization is 3 dB. But that is true only if the two polarizations are perfect. In fact, polarization mismatch loss can range to infinity (cross polarized).

polarization has a cross polarization counter point on the opposite side of the sphere.[52]

Many text books do include a discussion of the Poincare' sphere. So, we cannot fault them too much for short changing us on the polarization ellipse. But this is as far as they take you. You still must operate with spherical trigonometry.

See Appendix D on how to create your own polarization

See Appendix E on yet another way to determine polarization.

---

[52] In previous sections we have resolved vectors into vertical and horizontal components. These are cross polarized, meaning that they are independent of each other and do not influence each other. Indeed, that is why they are chosen for analysis.

I have invented a way to solve the equations using Cartesian trigonometry. So far as I know no one else has proposed this method.

My Paper on Polarization

Simplified Poincare' Sphere in Cartesian Coordinates with Dot Product

Eldon Gordon

14 May 2008

This paper defines the methodology and equations of the Poincare' sphere in terms of Cartesian coordinates. This methodology greatly simplifies the equations of polarization. In addition, this treatment is much more intuitive and promotes a simplified way to view the principles of electromagnetic polarization as opposed to the more complicated spherical coordinate equations. An important and unique feature of this methodology is the use of the dot product in the solution of the polarization equations.

On the Poincare' sphere the two polarizations that occur on opposite sides of any diameter line are orthogonal. Any desired electromagnetic polarization can be created by vector addition of the orthogonal components of the pair.

Henri Poincare' constructed the Poincare' sphere in the 1890s. It is a mathematical tool in the same manner that a nomogram is a mathematical tool except that unlike a nomogram the Poincare' sphere possesses exact mathematical equations.

The "surface" of the Poincare' sphere has a value of 1.0.

The Poincare' Sphere

Poincare' Sphere in xyz Format

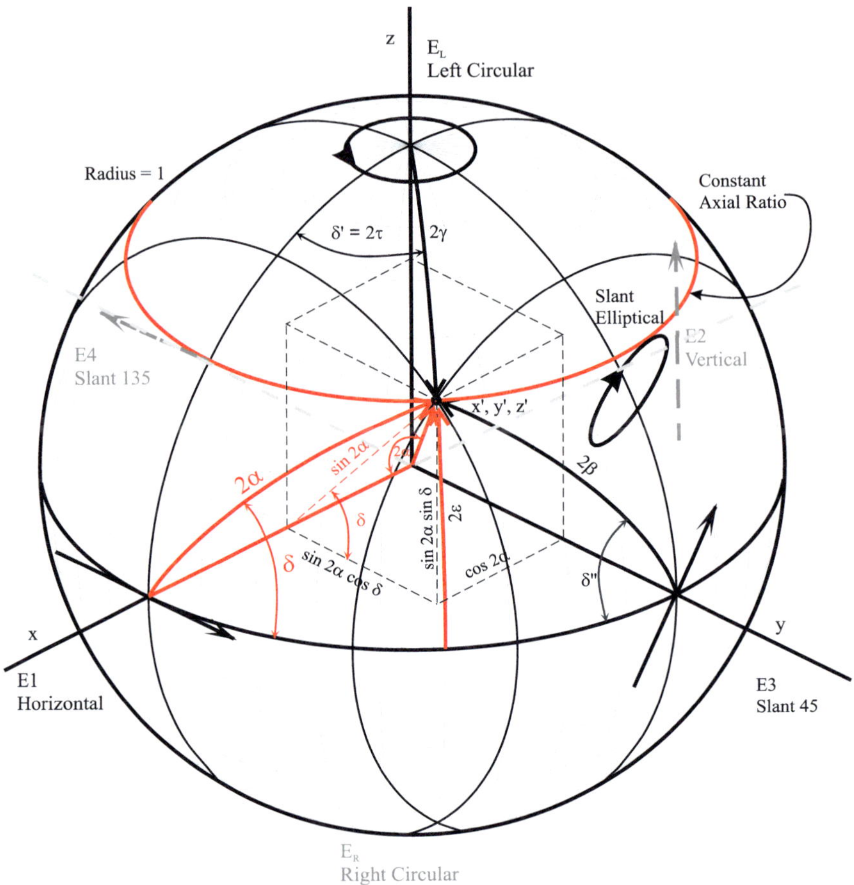

This view of the Poincare' sphere shows the polar geometry of orthogonal pairs:

The Horizontal/Vertical pair
The Slant 45/Slant 135 pair
The Left Handed Circular/Right Handed Circular pair

Any of the pairs can be similarly used to express polarization values, but the Cartesian coordinate equations are different for each pair.

The x,y,z Poincare' Sphere

This figure shows the Poincare' sphere for the V/H configuration. The red lines show the derivations of the equations. Almost all antenna measurement ranges use the vertical/horizontal method.

Poincare Sphere for V/H Polarization Orientation

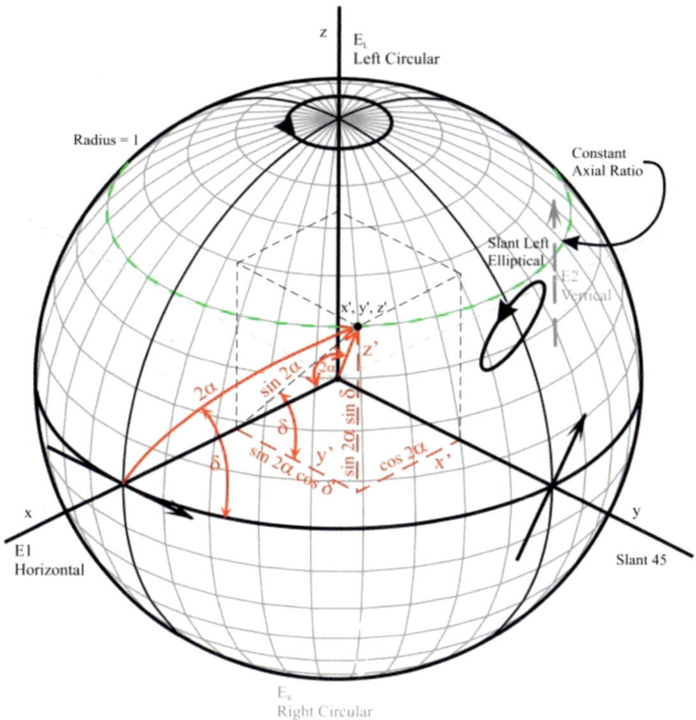

$E = 10^{\frac{P_x}{20}}$ Power to voltage

$\alpha = \tan^{-1}\frac{E_2}{E_1} \qquad 0 \geq \alpha \geq 90^\circ$

$\delta = \phi_{E_2} - \phi_{E_1} \qquad 0 \geq \delta \geq 360^\circ$

$x' = \cos 2\alpha$

$y' = \sin 2\alpha \cos\delta$

$z' = \sin 2\alpha \sin\delta$

$\varepsilon = \frac{\sin^{-1} z'}{2}$

$\text{Tilt} = \frac{\tan^{-1}(y'/x')}{2}$

Companion Antenna	x	y	z
RHCP	0	0	-1
LHCP	0	0	+1
Vertical	-1	0	0
Horizontal	+1	0	0

$E_T = \sqrt{E_2^2 + E_1^2} = \sqrt{E_{max}^2 + E_{min}^2}$

$E_{max} = E_T \cos\varepsilon$

$E_{min} = E_T \sin\varepsilon$

$\text{Axial Ratio} = \frac{E_{max}}{E_{min}} = \frac{1}{\tan\varepsilon}$

$\cos 2\theta = xx' + yy' + zz' = (x,y,z)\cdot(x',y',z')$ *(Dot Product)*

$\theta = \frac{\cos^{-1}(xx' + yy' + zz')}{2}$

$\Gamma = \cos^2\theta$; (Efficiency)

$P_L = 10\log(\Gamma)$ (Polarization Loss in dB)

$P_R = \Gamma(E_2^2 + E_1^2) = \Gamma(E_{max}^2 + E_{min}^2)$

(Received Power)

The method works from two sets of x,y,z coordinates. One set is the subject antenna polarization and the other set is the companion polarization coordinates. The source antenna is almost always linear or circular. The coordinates for these is given in the "Companion Antenna" block and is given as x,y,z. The polarization of the antenna under test has been determined by measurement in a test range and is given by x',y',z'. Given these two sets you can determine all of the polarization parameters of the unit-under-test as well as the polarization mismatch loss. The next two pages show the equations that I wrote in MathCAD. The great thing is that these are function equations and you can put a data matrix in arguments of the functions. So, you can get a tabulation of data very quickly. Note that I used "p" for the measured antenna and "w" for the companion antenna. Also, "A" is measured amplitude in dB and "P" is measured phase in degrees.[53]

---

[53] Notice that the computation of a measured polarization requires both calibrated amplitude and calibrated phase. This is okay for most indoor measurement ranges. But outdoors the antenna mounting platforms tend to move in the wind, making accurate phase measurements difficult. See Appendix F for a method to measure polarization without phase.

Equations of the X, Y, Z Poincare' Sphere

Copyright Eldon Gordon

Assumes input data: gain (dB) of A2, A1 and phase (degrees) of P2, P1
Gain values are relative, subject to range calibration

Polarization Voltages

$$E2(A2) \equiv 10^{\frac{A2}{20}}$$

$$E1(A1) \equiv 10^{\frac{A1}{20}} \qquad E_T(A2,A1) \equiv \sqrt{E2(A2)^2 + E1(A1)^2}$$

Alpha

$$\alpha(A2,A1) \equiv \operatorname{atan}\left(\frac{E2(A2)}{E1(A1)}\right)$$

Delta

$$\delta(P2,P1) \equiv (P2\cdot deg - P1\cdot deg)$$

The x, y, z components on the Poincare' sphere

$$xp(A2,A1) \equiv \cos(2\cdot\alpha(A2,A1))$$

$$yp(A2,A1,P2,P1) \equiv \sin(2\cdot\alpha(A2,A1))\cdot\cos(\delta(P2,P1))$$

$$zp(A2,A1,P2,P1) \equiv \sin(2\cdot\alpha(A2,A1))\cdot\sin(\delta(P2,P1))$$

Separation Angle (θ)

$$Theta(A2,A1,P2,P1,xw,yw,zw) \equiv \frac{\operatorname{acos}(xw\cdot xp(A2,A1) + yw\cdot yp(A2,A1,P2,P1) + zw\cdot zp(A2,A1,P2,P1))}{2}$$

Efficiency

$$\Gamma(A2,A1,P2,P1,xw,yw,zw) \equiv \cos(Theta(A2,A1,P2,P1,xw,yw,zw))^2$$

$$Eff(A2,A1,P2,P1,xw,yw,zw) \equiv \Gamma(A2,A1,P2,P1,xw,yw,zw)\cdot 100$$

Polarization Loss

$$PL(A2,A1,P2,P1,xw,yw,zw) \equiv 10\cdot\log(\Gamma(A2,A1,P2,P1,xw,yw,zw)) \quad \text{dB}$$

$$Pt(A2,A1) \equiv \left(E1(A1)^2 + E2(A2)^2\right) \quad \text{Total power in the wave}$$

Gain

$$Gain(A2,A1,P2,P1,xw,yw,zw) \equiv 10\cdot\log(\Gamma(A2,A1,P2,P1,xw,yw,zw)\cdot Pt(A2,A1))$$

ε, The angle associated with constant axial ratio

$$\varepsilon(A2,A1,P2,P1) \equiv \frac{\operatorname{asin}(zp(A2,A1,P2,P1))}{2}$$

Axial Ratio

$$AR(A2,A1,P2,P1) \equiv 20\cdot\log\left(\frac{1}{\operatorname{if}\left(\tan(\varepsilon(A2,A1,P2,P1)) = 0, 0.0001, \sqrt{\tan(\varepsilon(A2,A1,P2,P1))\cdot\tan(\varepsilon(A2,A1,P2,P1))}\right)}\right)$$

Maximum Gain

$$P_T(A2,A1) \equiv E1(A1)^2 + E2(A2)^2 \quad Pmax(A2,A1) \equiv 10\cdot\log(P_T(A2,A1))$$

Tilt

$$Tilt(A2,A1,P2,P1) \equiv \operatorname{atan}\left(\frac{yp(A2,A1,P2,P1)}{xp(A2,A1)}\right) \div 2 \div deg$$

Maximum and minimum Voltages

$$Emax(A2,A1,P2,P1) \equiv \sqrt{\left(E_T(A2,A1)\cdot\cos(\varepsilon(A2,A1,P2,P1))\right)\cdot E_T(A2,A1)\cdot\cos(\varepsilon(A2,A1,P2,P1))}$$

$$Emin(A2,A1,P2,P1) \equiv \sqrt{\left(E_T(A2,A1)\cdot\sin(\varepsilon(A2,A1,P2,P1))\right)\cdot E_T(A2,A1)\cdot\sin(\varepsilon(A2,A1,P2,P1))}$$

Here is an example for testing the function equations

$A2 \equiv -4 \quad A1 \equiv -2 \quad P2 \equiv -98 \quad P1 \equiv 37 \quad xw \equiv 0 \quad yw \equiv 0 \quad zw \equiv 1$

Use this sheet by lining it up on the right hand side of the previous sheet.

$E2(A2) = 0.631$

$E1(A1) = 0.794$

$\alpha(A2, A1) = 0.671$

$\delta(P2, P1) = -2.356$

$xp(A2, A1) = 0.226$

$yp(A2, A1, P2, P1) = -0.689$

$zp(A2, A1, P2, P1) = -0.689$

$\text{Theta}(A2, A1, P2, P1, xw, yw, zw) = 1.165$

$\Gamma(A2, A1, P2, P1, xw, yw, zw) = 0.156$

$\text{Eff}(A2, A1, P2, P1, xw, yw, zw) = 15.562$

$PL(A2, A1, P2, P1, xw, yw, zw) = -8.079$

$Pt(A2, A1) = 1.029$

$\text{Gain}(A2, A1, P2, P1, xw, yw, zw) = -7.955$

$\varepsilon(A2, A1, P2, P1) = -0.38$

$AR(A2, A1, P2, P1) = 7.974$

$\text{Pmax}(A2, A1) = 0.124$

$\text{Tilt}(A2, A1, P2, P1) = -35.907$

$\text{Emax}(A2, A1, P2, P1) = 0.021$

$\text{Emin}(A2, A1, P2, P1) = 5.391 \times 10^{-3}$

An example of how to use these equation with matrices of data. These are the "measured" values for the UUT

$$M \equiv \begin{pmatrix} 0 & 0 & 90 & 0 \\ 0 & -40 & 0 & 0 \\ 5 & 3 & -85 & 75 \\ 1 & 6 & 55 & 43 \\ -9 & -6 & 24 & -165 \end{pmatrix}$$

$$\overrightarrow{\text{Gain}\left(M^{\langle 0\rangle}, M^{\langle 1\rangle}, M^{\langle 2\rangle}, M^{\langle 3\rangle}, -1, 0, 0\right)} = \begin{pmatrix} 0 \\ 0 \\ 5 \\ 1 \\ -9 \end{pmatrix}$$

Equations in Excel

P2		P1			
0.00		0.50		Degrees to Radians	
E2	A2	E1	A1	0.017453293	
1.00	-21.07	1.06	52.69		
Alpha	Delta				
43.35	-73.76				
Measured Polarization			Companion Antenna Polarization		
xp	yp	zp	xa	ya	za
0.06	0.28	-0.80	0.00	0.00	-1.00
Eliptic(e)	(in degrees)				
-0.46	-26.57			OK	
Tilt	(in degrees)				
0.68	39.18				
Theta	(in degrees)				
0.32	18.43				
Efficiency	Polarization Loss		Gain (dB)	Axial Ratio	
0.900	0.458	dB	2.810	6.02	dB

This graphic and the equations above apply to measurements taken from two antennas, one vertical and one horizontal. The two antennas must be calibrated to equal gain and phase.

When the transmitting and receiving antennas have different polarizations;i.e., one circular and one elliptical, there is a loss of received power due to polarization mismatch.

The Poincare sphere is a mathematical tool in the same manner as a nomogram. Any polarization can be positioned on the Poincare sphere, radius = 1. The polarization mismatch loss is computed between the angular distance on the Poincare sphere (2θ) between the transmit antenna polarization and the receive antenna polarization. When 2θ = 180 degrees such as from vertical to horizontal, the antennas are "cross polarized" and there is zero power transfer.

On this worksheet the two polarizations are expressed in terms of the x,y,z components of the points on the Poincare sphere.

In this example, the companion antenna is RHCP.

I have all of the above posted at www.barnswallowpublications.net . You are free to download the Excel file.

Although all the above is copyrighted you are free to use it in your own endeavors, even if you are working for a company. But, if you intend to use it in a publication for sale, you need my permission.

Radomes

A radome is a cover to protect an antenna from the environment. My work has primarily been with communications satellite antennas on the top of an aircraft fuselage. With such an antenna there is one thing that is certain.

A radome distorts (degrades) polarization.

The antennas are usually, but not always, circularly polarized. But after the signal passes through the radome it is elliptically polarized. The space based communications satellite antenna is circularly polarized. So, there is a polarization mismatch between the two.

Another factor is that satellite adjacent channels have cross polarized antennas for isolation, one RHCP and the other LHCP. In perfect world these two would completely reject each other. But when the signal leaving the aircraft is elliptically polarized, both antennas will receive some signal. A primary objective of radome design is to ensure that the cross polarization between RHCP and LHCP remains high. The equations I have will compute that value[54] as well as all of the other parameters associated with the radome effects.

An in-depth description of radome operation is beyond the scope of this book. But we can get the general flavor by considering the simple flat plate radome.

---

[54] The equations to determine the polarization distortion is quite complex and not many people have them; not even radome manufacturers.

The Most Simple Radome

Vectorial effect at each substrate layer of the radome

These rays sum vectorially to form a single transmitted signal. The rays exiting the radome are always parallel to the incident ray regardless of the construction of the radome.

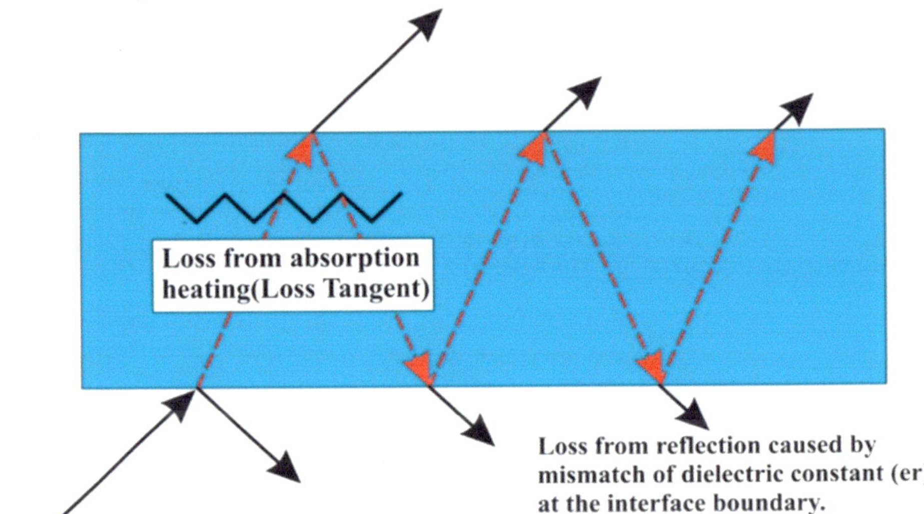

Incident ray **These rays sum vectorially to form a single reflected signal. When the thickness is one-half wavelength in the material the sum becomes zero and the radome is "invisible", except for loss.**

This occurs at every interfacing layer of the radome, so the overall effect is very complex and the equations to solve the net result are also very complex. Also, the two polarizations comprising circular polarization pass through differently, thereby distorting the circular polarization to elliptical polarization.

Here is another representation showing the equations.

Let Z1 = Z2 = 377 Ohms and let Zo be the impedance of the slab. Then the is direct correspondence to these two figures.

This is a more general set of equations for the case of three stacked impedances.

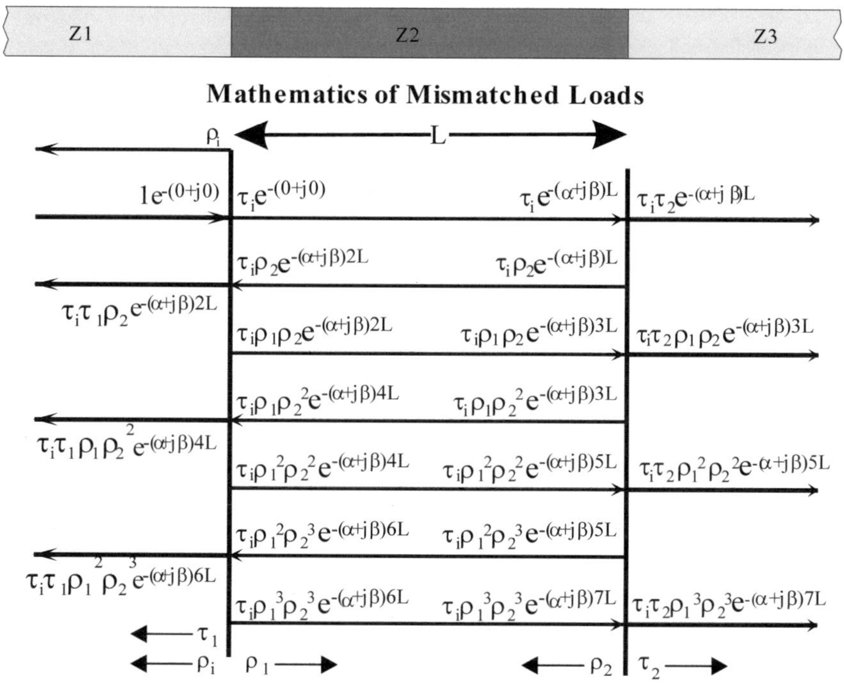

Start by defining ρ_1 ρ_2 τ_i, τ_1, τ_2

Note that τ_i associates with ρ_i not ρ_1.

Reflection Coefficient

$$R = \rho_i + \tau_i\tau_1 \sum_{n=0}^{\infty} \rho_1^n \rho_2^{n+1} e^{-(\alpha+j\beta)2L(n+1)}$$

In closed form

$$R = \rho_i + \frac{\tau_i\tau_1 \rho_2 e^{-(\alpha+j\beta)2L}}{1 - \rho_1 \rho_2 e^{-(\alpha+j\beta)2L}}$$

$$\rho = \frac{Z_L - Z_0}{Z_L + Z_0}$$
$$\tau = \frac{2Z_L}{Z_L + Z_0}$$
$$\tau = 1 + \rho$$

Transmission Coeffecient

$$T = \tau_i\tau_2 \sum_{n=0}^{\infty} \rho_1^n \rho_2^n \; e^{-(\alpha+j\beta)L(2n+1)}$$

In closed form

$$T = \frac{\tau_i\tau_2 e^{-(\alpha+j\beta)L}}{1 - \rho_1\rho_2 e^{-(\alpha+j\beta)2L}}$$

In the equations α is the loss component and β is the phase component.

This figure shows a transmission line instead of a dielectric slab. But, here is something to remember. **Slabs in space and physical**

transmission lines are governed by the same equations.

In both cases, I am showing a single layer. Most radomes have at least three layers. Then, the equations get very complex.

Here is a plot of loss in the slab versus frequency.

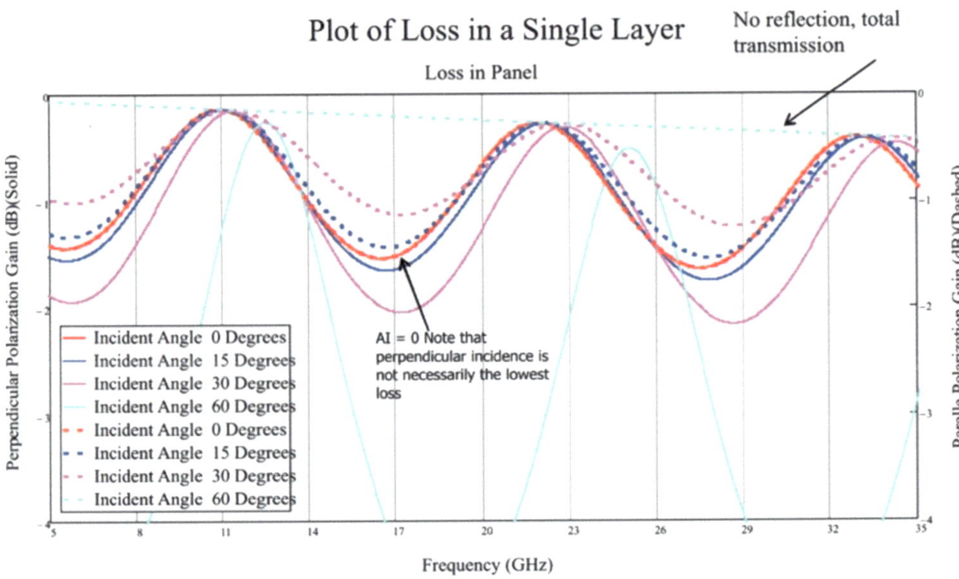

This graphic shows the responses in single layer. At the peaks, reflections are zero and the remaining loss is all absorptive. Note that at IA = 0 degrees, perpendicular loss and parallel lose are identical (solid overlaying dashed). The parallel loss at 60 degrees is all absorptive because of the Brewster angle effect. **The peaks of different incident angles (IA) occur at different frequencies.**

This graphic shows an antenna under a radome and emphasizes the polarization distortion. I sometimes call it “polarization twisting”.

Here is a graphic that is great help to visualize how the ray transits through the radome.

Radome Geometry

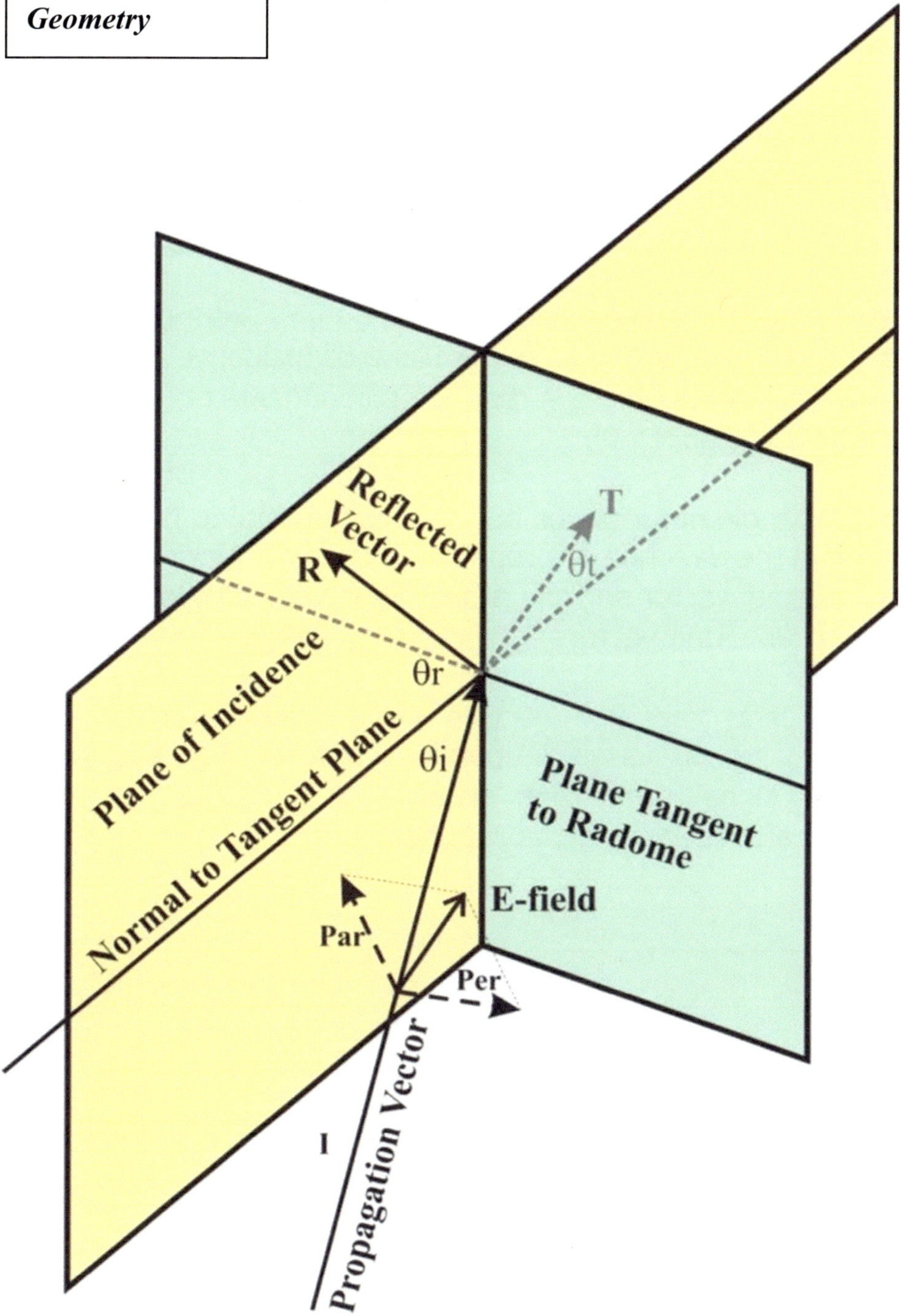

Everywhere a perpendicular ray from the antenna intercepts the radome there is a tangent plane and a Plane -of-Incidence

Vectors I, R, T, Par and Normal lie in the Plane -of-Incidence,
Vector Per is perpendicular to the Plane of Incidence
Vectors Par and Per are the resolved components of the E-field vector.

θi is the angle of incidence between the normal to the tangent plane and the direction of propagation vector The dot product of the normal vector and the propagation vector yields the cosine of θi. (Unit vectors)

The Par and the Per vectors pass through the radome differently, as discussed in previous slides. So, a circularly polarized signal from the antenna will not be circularly polarized after transiting the radome.

A Radome

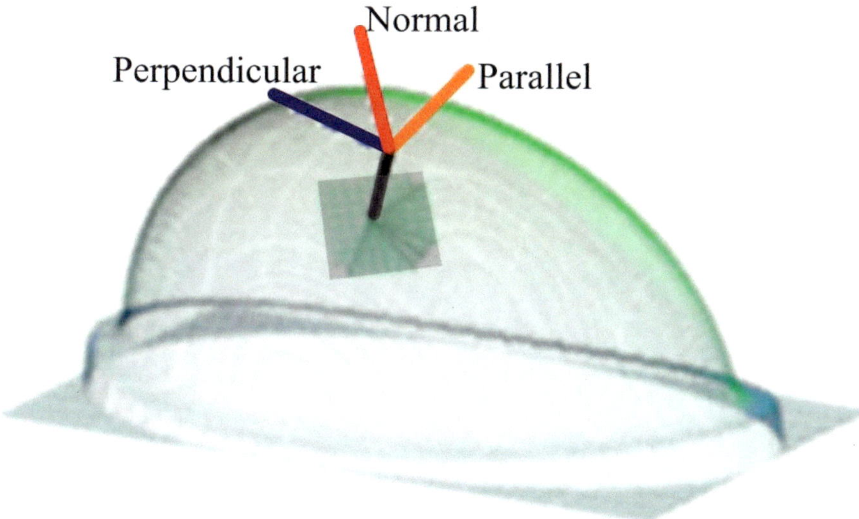

Here is an example of the radomes that I have designed and evaluated. Let your brain flip until the closer side is down. Note that the perpendicular and parallel components are not vertical/horizontal. But the analysis software must mathematically rotate them to vertical/horizontal so the summation in the far-field can be accomplished. Every point on the radome requires a different rotation.

Elevation from zenith is $\theta = 45$ degrees. Azimuth from rear is $\phi = 45$ degrees
Pointing error $PEC = 0.13$ degrees Beamwidth $BW = 4.81$ degrees
Frequency of this evaluation is $freq = 14.2 \cdot GHz$. Highest Sidelobe $HSL = -23$ dB

$LossV = 0.1$ dB $PhsV = -32.9$ deg $LossVcros = 37.1$ dB $PhsVcros = 55.1$ deg
$LossH = 0.11$ dB $PhsH = -33.6$ deg $LossHcros = 37.1$ dB $PhsHcros = -124.9$ deg
$LossCP = 0.1$ dB Axial Ratio $ART = 0.09$ dB CP Isol $CalCP(ART) = ">45"$ dB

The equations to compute these parameters are quite complex. Not very many people have them, not even some radome manufacturers.

Appendix A

Derivation of Velocity and Acceleration

Finding the values of velocity and acceleration to use in the Lamor formula

Equation of Radiation by an accelerating charge defined J. J. Larmor in 1897. Example:

Electron Charge

Electron Acceleration (Assumed value to produce 1 watt)

$$e \equiv 1.602176565 \cdot 10^{-19} \cdot \text{coul} \quad a \equiv 5 \cdot 10^{6} \cdot \frac{\text{m}}{\text{sec}^2}$$

Density of free electrons in copper

$$\text{Den} \equiv 8.49 \cdot 10^{28} \cdot \frac{1}{\text{m}^3}$$

Let a wire be 12 inches long and 0.02 inches in diameter

$$L \equiv 12 \cdot \text{in} = 0.305 \cdot \text{m}$$

$$r \equiv 0.01 \cdot \text{in} = 2.54 \times 10^{-4} \cdot \text{m}$$

Volume assuming 2 micrometer skin depth

$$\text{Vol} \equiv \pi \cdot r^2 \cdot L \cdot 0.016 = 9.884 \times 10^{-10} \cdot \text{m}^3$$

Number of electrons in the volume

$$N \equiv \text{Den} \cdot \text{Vol} = 8.392 \times 10^{19}$$

Select the acceleration required to produce 1.0 watt (peak) by using the Larmor formula.

$$P \equiv \frac{(N \cdot e)^2 \cdot a^2}{6\pi \cdot \varepsilon_0 \cdot c^3} = 1 \cdot \text{watt}$$

Let the period be 1 ns.

$i \equiv 0 \ldots 100$

$$Freq \equiv \frac{1}{1 \cdot ns} = 1 \cdot GHz$$

$$t_i \equiv 1 \cdot ns \cdot \left(\frac{i}{100} + 10^{-20} \right)$$

Select a value of v to produce a peak acceleration of a.

$$v \equiv 7.963 \cdot 10^{-4} \cdot \frac{m}{s}$$

$$a = 5 \times 10^6 \cdot \frac{m}{s^2}$$

Differentiate velocity to get acceleration

$$AL(v) \equiv \left| \begin{array}{l} \text{for } \; i \in 0 .. 10 \\ \quad A_i \leftarrow \dfrac{v \cdot \sin\left(2\pi \cdot Freq \cdot t_{i+1}\right) - v \cdot \sin\left(2\pi \cdot Freq \cdot t_i\right)}{t_{i+1} - t_i} \\ A \end{array} \right.$$

$$a = 5 \times 10^6 \cdot \frac{m}{s^2}$$

$$\max(AL(v)) = 5 \times 10^6 \cdot \frac{m}{s^2}$$

Wavelength is:

$$\lambda \equiv \frac{c}{Freq} = 11.803 \cdot in$$

Plot acceleration m/s^2

Plot velocity m/s

$Accel(t) \equiv a \cdot \cos(2\pi \cdot Freq \cdot t)$

$Vel(t) \equiv v \cdot \sin(2\pi \cdot Freq \cdot t)$

$BB \equiv Accel(t)$

$AA \equiv Vel(t) \cdot 1000$

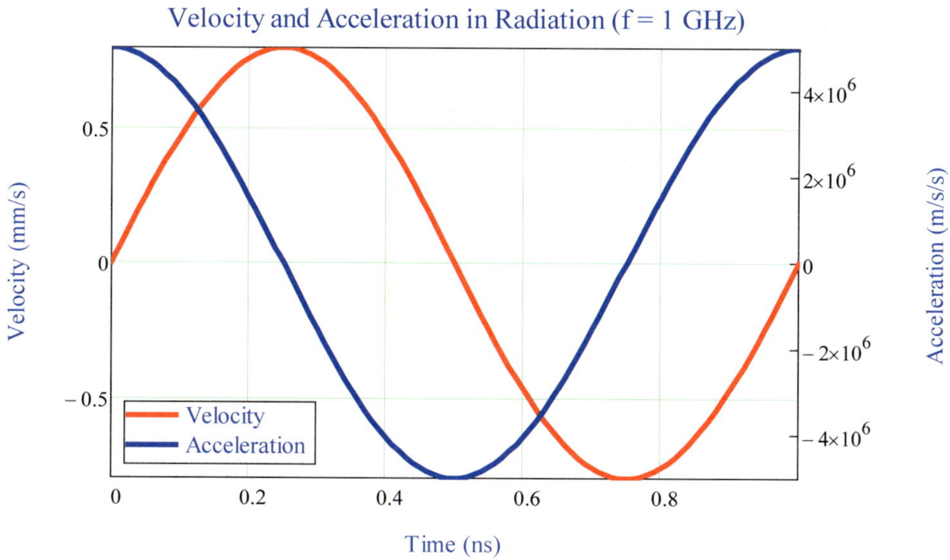
Velocity and Acceleration in Radiation (f = 1 GHz)
Velocity (mm/s)
Acceleration (m/s/s)
Time (ns)
0.5
0
– 0.5
4×10^6
2×10^6
0
– 2×10^6
– 4×10^6
0
0.2
0.4
0.6
0.8
Velocity
Acceleration

Appendix B

Complex Impedance and Circuits

Complex impedance applies to alternating current and EM waves. It is characterized by having two components, real and imaginary. Take Z = R + jX where Z is the complex impedance; R is the real part and X is the imaginary part. All have the dimension of Ohms.

When I started writing this book, I thought that I would not need to research the very basic stuff. But then I looked at "j" and said to myself, "Where the heck did that come from?" After all it is just the square root of minus one." So, I researched it and found that many people had the same question. If you are curious, I suggest you do the same. But for now, I accept it as one of the miracles I referenced in the first chapter. What the "j" does is allow us to utilize Eulers equation to treat sine waves of different phases as if they were vectors, which they are not.

Many applications of complex numbers require that we use conjugate multiplication. The conjugate of complex number is the same number except the sign of the imaginary part is changed.

Example:
$(a + jb)^2 = (a + jb)* (a + jb) = a^2 + j2ab - b^2$; a complex number

And $(a + jb)* (a - jb) = a^2 + b^2$; a real number

They both have the same absolute value, but in some cases the presence of the imaginary part messes us up.

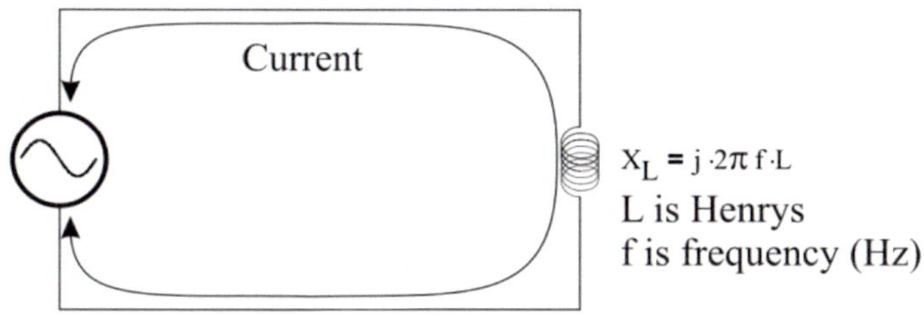
Current
X_L = j·2π f·L
L is Henrys
f is frequency (Hz)

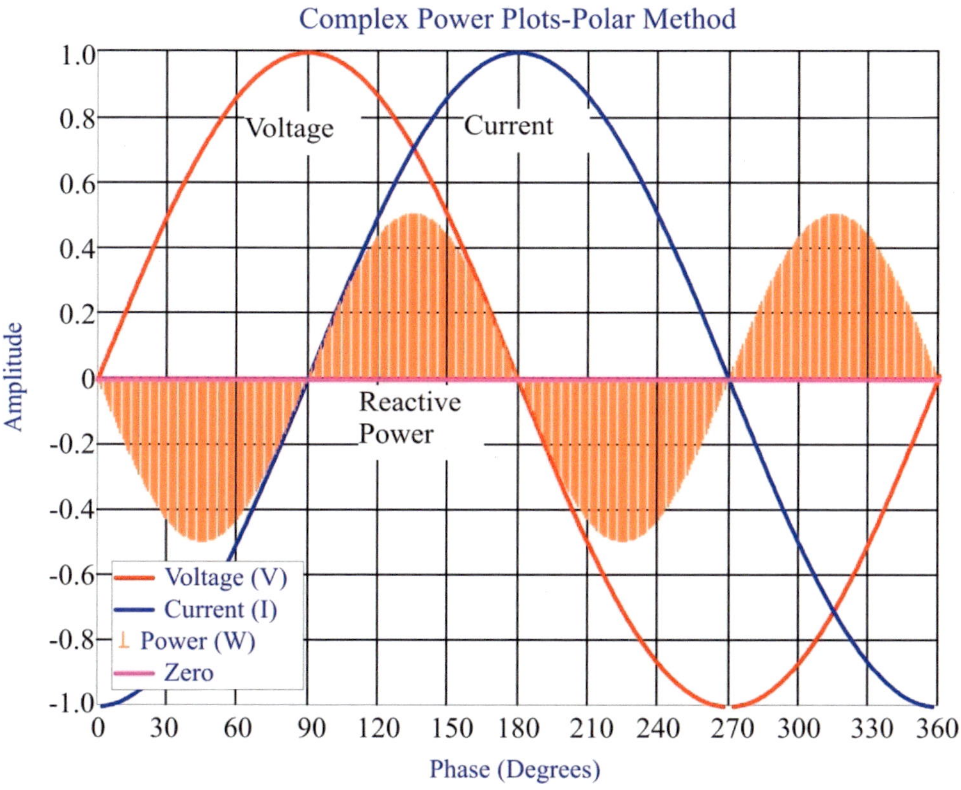
Complex Power Plots-Polar Method
Voltage
Current
Reactive Power
Amplitude
Phase (Degrees)
Voltage (V)
Current (I)
Power (W)
Zero

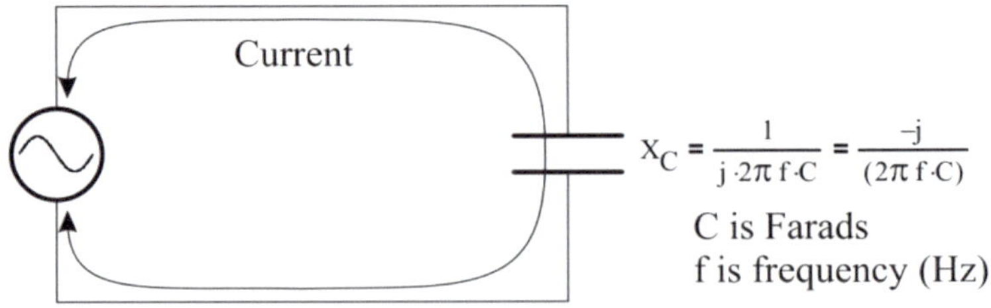

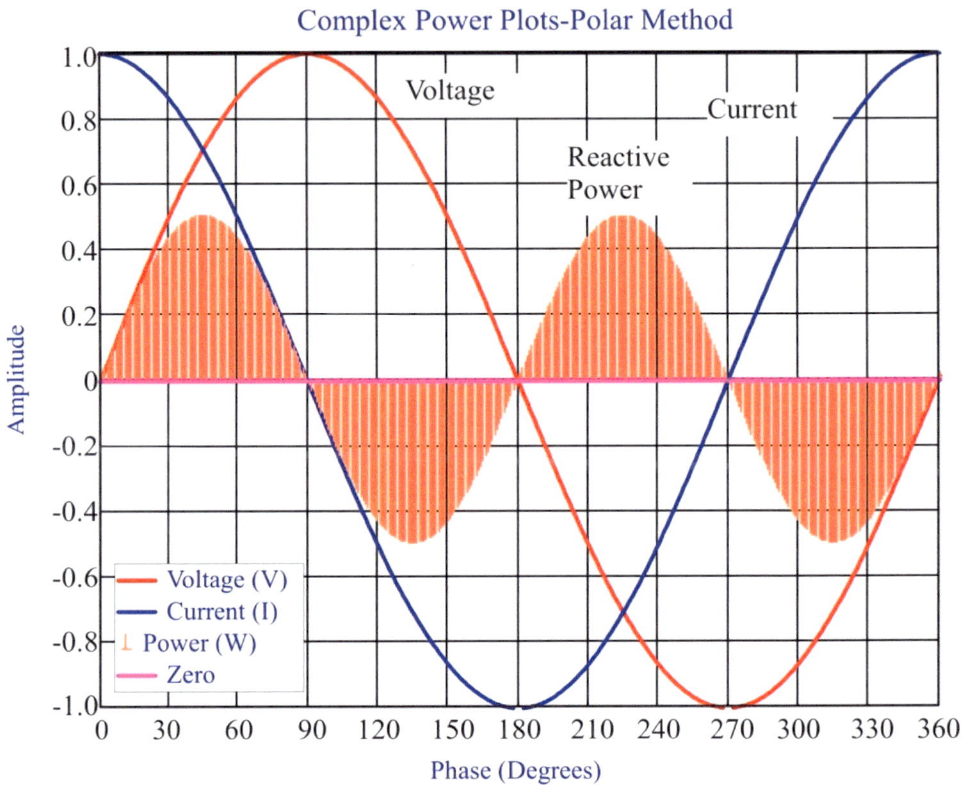

You can see that in each case the voltage and current have a phase difference of 90 degrees. The reactive power plot shows that, in each half cycle, the reactance receives energy from the source, then returns it to the source---net zero power. So, with these lossless components, no heat is generated.

But, in truth, both inductors and capacitors have some loss and will generate some heat. The voltage and current will be not quite 90 degrees apart.

Also, note that the reactance value is a function of frequency. Inductive reactance is positive and capacitive reactance is negative (because 1/j = -j).

Now observe the polar plot with pure resistance.

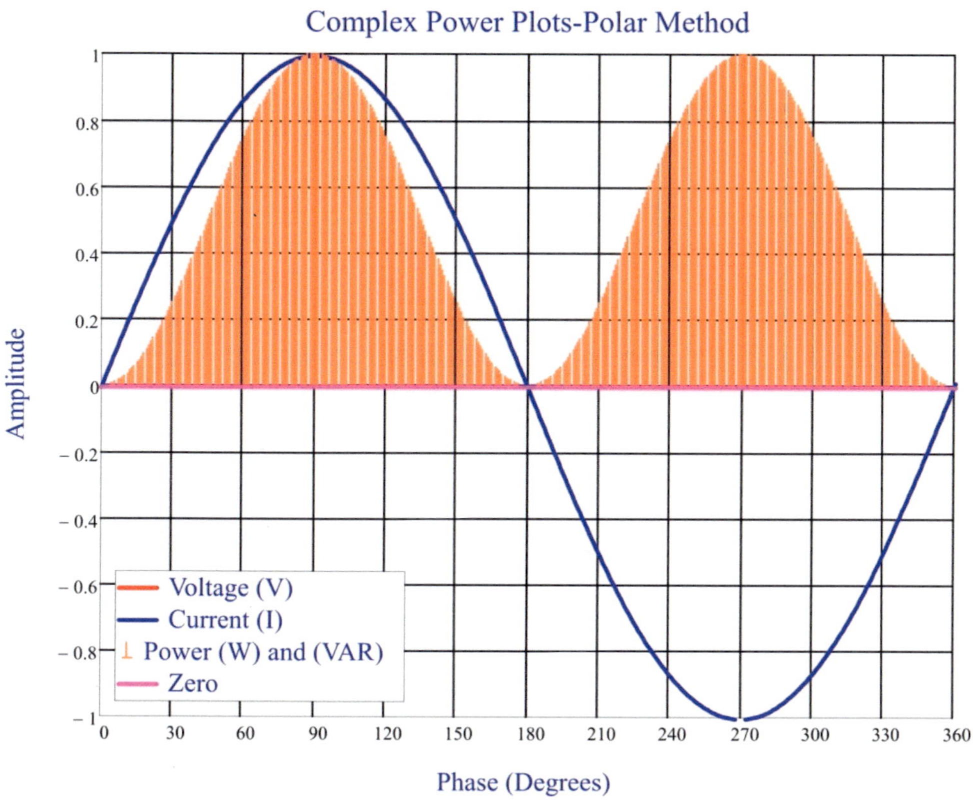

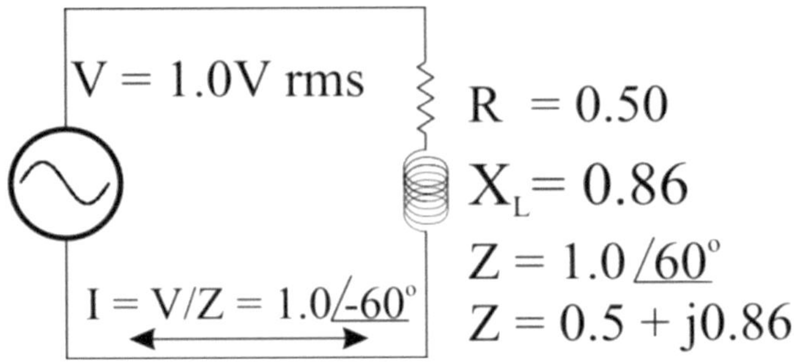

Power is dissipated only in the "R" component. The "R" does not have to be a resistor. It can stand in for anything that is using power including electric motors, wiring losses, dielectric losses, etc. The loss of the reactive component(s) can be rolled into "R". If desired, multiple components can be mathematically combined into a single resistance and a single reactance. The graphic is shown with an inductor, but it can be modeled as a capacitor by simply changing the sign of the imaginary part from plus to minus. In this representation of a circuit, the current is the same everywhere. The value of the current is still 1A, but the phase will be different from the voltage phase.

Two ways to analyze

Notice that the total impedance "Z" is shown two ways.

Polar Method

The first "Z" equation is a polar type. It simply states that the impedance has an absolute value of 1.0 ohms and a phase angle of 60 degrees. The current is V/Z or 1A rms with phase 60 degree difference from the voltage phase. Mathematically, this is Power = V times I or;

Power = sin(ωt)∗sin(ωt + 60°) . Notice that there is no imaginary component, "j". The power and motor industries tend to use this representation. (Before solving this equation, the 60° must be converted to radians, unless your computation software does it automatically.) The "ωt" is the time phase of the voltage and current

The plots shown above are computed by the polar method.[55]

The Rectangular Method

Now I = V/(R + jX) and Power = V∗I. (Power is also equal to V^2/Z and $I^2 * Z$)[56]

Defining Power Terms

In a complex circuit energy moves from the source and is stored in the reactive component. In the next half cycle this stored energy is returned to the source. The power types are defined thus:
Apparent Power = V∗I and is called VA (Volt Amp)
Actual Power is the power that does work and is V∗I∗cos(θ) where θ is the phase difference between the voltage and the current and is called watts.

---

[55] I use MathCAD in my work. MathCAD cannot solve polar equations directly. I had to create equations to do it.

[56] Actually, V^2 is computed as V times the conjugate of V. If you simply square V, you get an imaginary component that may cause an error in the result. The same is true for I^2. You see this often in electromagnetic text books.

$V*I*\sin(\theta)$ is the "power" that is circulating between the reactive component and the source and is called VAR (Volt Amp Reactive)
You will recognize these as the real and imaginary parts of Euler's equation.

Here are the illustrative plots as applied to the circuit shown above.

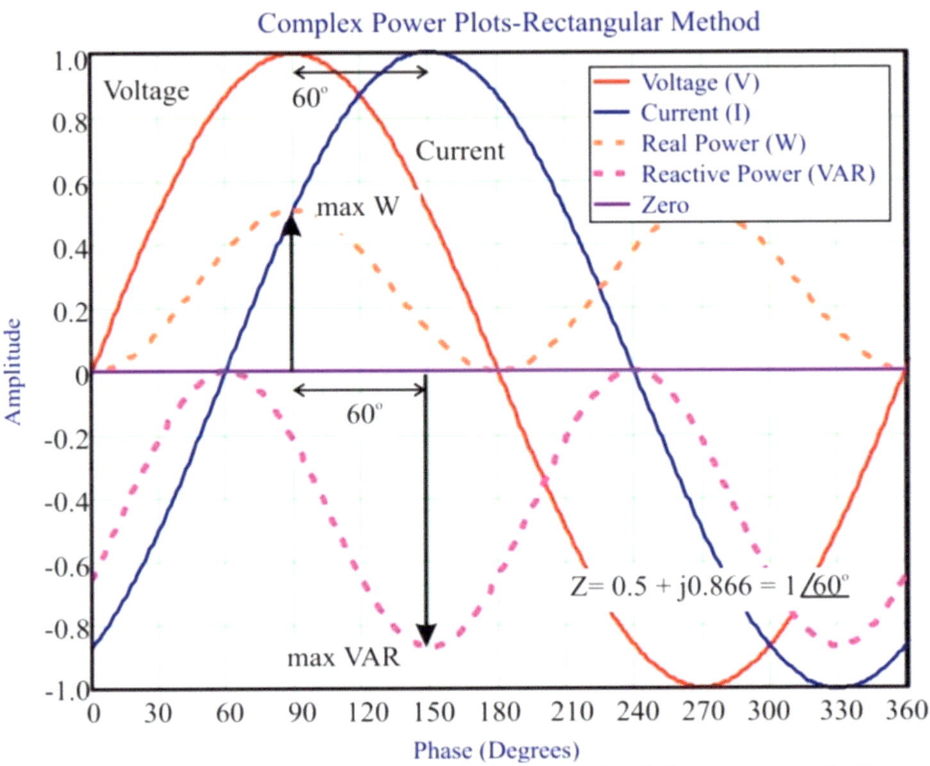

The Actual Power is 0.5 watts and Reactive "Power" is 0.866 VAR. Computed with rectangular parameter Power is equal to $V*I = P + jQ$; where P (the real part) is the actual power and Q (the imaginary part) is the VAR power. Observe that the real power 0.5 watts is one-half the power when the load was purely resistive (1.0 ohm and 1.0 watts)

The VAR plot is below the zero line for an inductive component and above the zero line for a capacitive component. The power plot (W) is always above the zero line.

Here is the plot by the polar method.

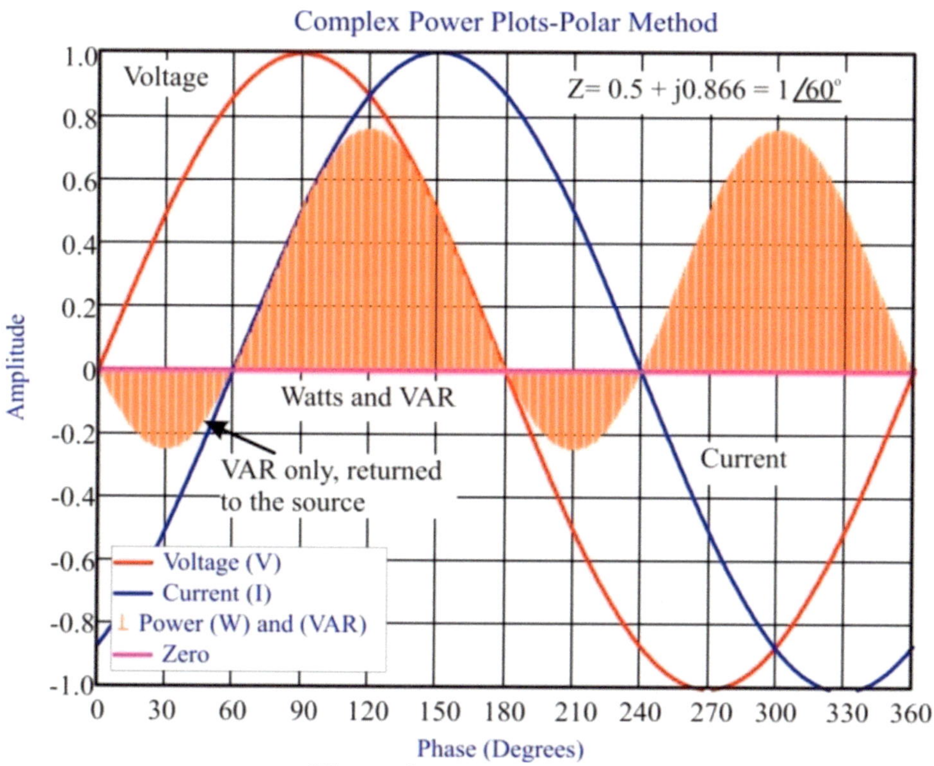

There is no observable way to determine the power (W) and reactive part (VAR). So, we have to resort to calculus to integrate under to curve to find the power.

$$\frac{\int_0^{\pi} \sin(w)^2 + \min(\mathrm{Pwr}(R, X))\, dw}{\int_0^{\pi} \sin(w)^2\, dw} \cdot 100 = 50$$

Here the absolute value of the current is always one and the phase is stepped from 0 to 90 degrees. The second term in the numerator is the peak of the value under the zero line. For our reference circuit (60°) the power is decreased by 50 percent; the same as we observed on the rectangular plot.

This plot shows the decrease in real power for a difference phase from 0 to 90 degrees.

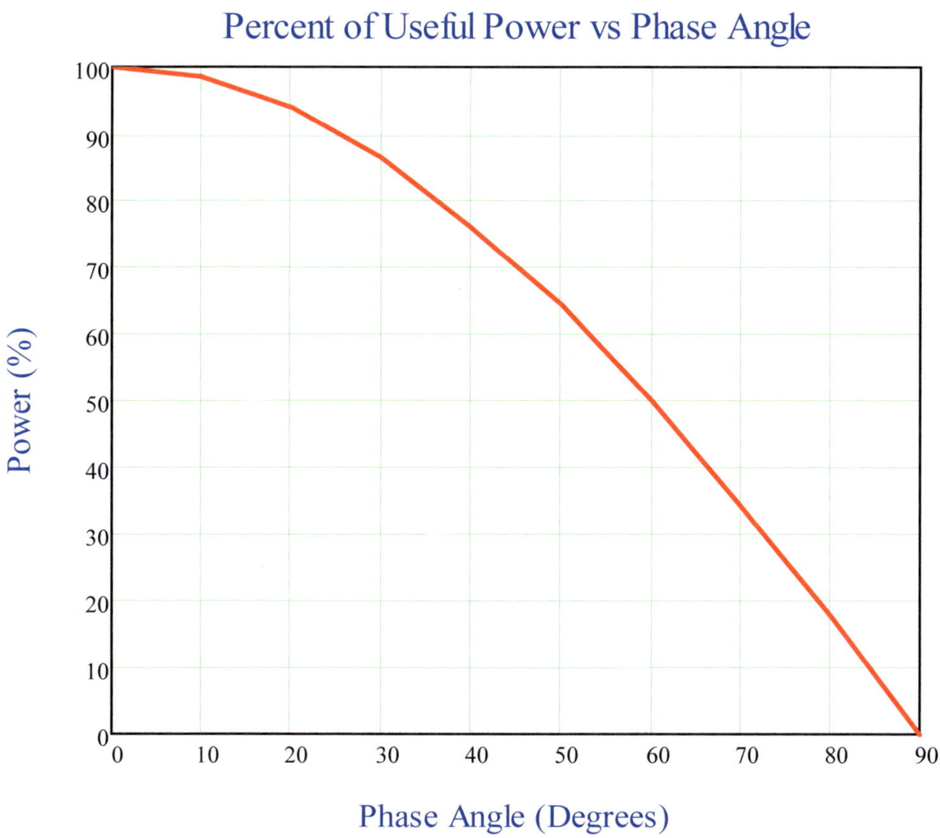
Percent of Useful Power vs Phase Angle
Power (%)
100
90
80
70
60
50
40
30
20
10
0
0
10
20
30
40
50
60
70
80
90
Phase Angle (Degrees)

Dual Reactance Components

Now let's look at two circuits; series and parallel; each consists of an alternating source and three components: a resistor, an inductor and a capacitor. The resistor (R) can be an actual resistor or an equivalence accounting for the losses in L and C.

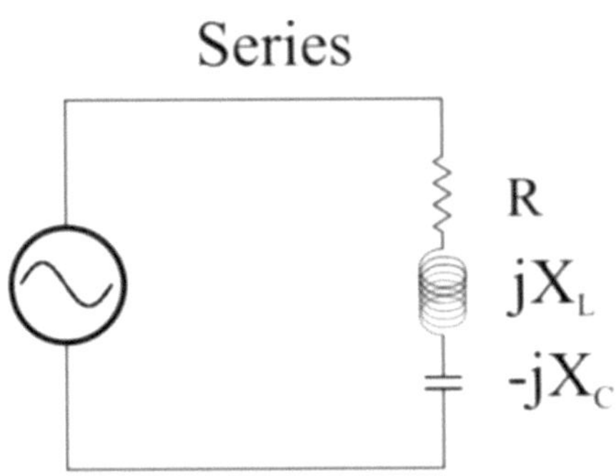

$$Z = R + jX_L - jX_C$$

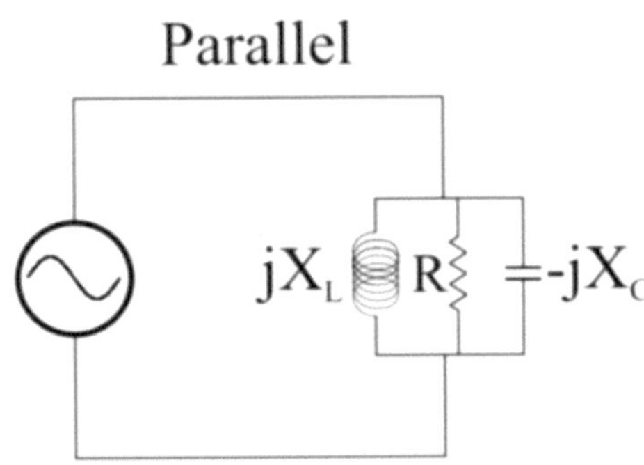

$$Z = \frac{1}{\frac{1}{R} + \frac{1}{jX_L} - \frac{1}{jX_C}}$$

Consider the case where $X_L = X_C$. The circuit is resonant at the frequency where the absolute magnitude of X_L and X_C are equal. At resonance the series L and C has very low impedance. The parallel L and C has a very high impedance. In both circuits the resulting total impedance is R. If R is not physically in the circuit it will be represented by the losses in L and C. For low components, the series circuit current is very high. In the parallel circuit the current is very low. Resonance occurs when $2\pi fL = 1/2\pi fC$; or $f = (1/(2\pi\sqrt{(L*C)})$

Maximum Power Transfer

Sometimes it is an objective get all power possible from a circuit. An obvious example is to get as much power as possible from an antenna into a receiver.[57]

Consider the following circuit
Z_S is the impedance looking back into the source. It includes the impedance of the source itself as well as any impedance in the connecting circuitry. Z_L is the impedance of the load. The only useful power is that which is consumed in the resistance, R_L. Maximum power transfer occurs when Z_S and Z_L are complex conjugates of each other.

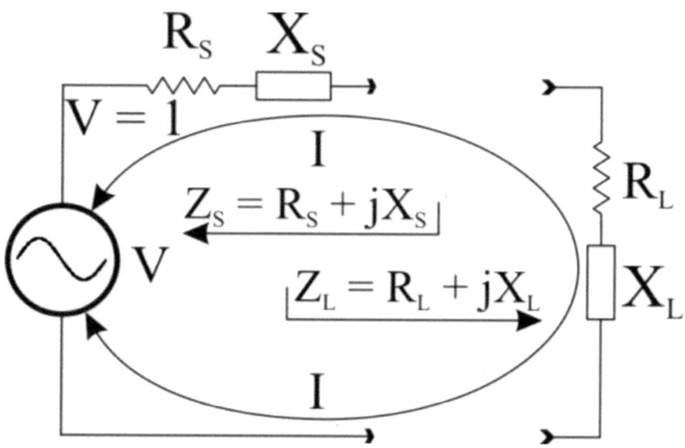

[57] There are many instances where maximum power transfer is not an objective. You just want enough power to run your device. Example: The battery in your car has a very low impedance; i.e., nearly a short circuit. Place a short across it and stand back watch the battery catch fire and burn up.

Here are the equations:

Maximum Power Transfer

$RS \equiv 1 \qquad XS \equiv 1.5 \qquad RL \equiv 1 \qquad XL \equiv -1.5$

$$IC(RS, XS, RL, XL) \equiv \frac{1}{RS + j \cdot XS + RL + j \cdot XL}$$

$IC(RS, XS, RL, XL) = 0.5$

Power in RL

$$Power(RS, XS, RL, XL) \equiv IC(RS, XS, RL, XL) \cdot \overline{IC(RS, XS, RL, XL)} \cdot RL$$

$Power(RS, XS, RL, XL) = 0.25$

The power delivered to R_L is 0.25 watts. Change any value such that Z_L is not the complex conjugate of Z_S and the power is less than 0.25 watts.

Appendix C

Lossy Transmission Lines

The equation that includes the loss of the transmission line is the similar to the lossless case except the tangent function is now a hyperbolic tangent with the argument ($\alpha + j\beta$), where α is the attenuation constant in Nepers/meter and β is the phase constant in radians/meter. **(Hint: to convert Nepers/meter to dB/meter, multiply by 20*log(e) = 8.686.)**

The following example is a typical coaxial transmission line operating at 1 GHz. Both with loss and without loss conditions are shown. The loss is set to 0.3 dB/meter. You can see that 100 feet of the cable has 9.1 dB of loss. Also, note that the input impedance is approaching 50 ohms, the characteristic impedance of the transmission line. This illustrates an earlier point, that a very long transmission line will have an input impedance equal to its characteristic impedance no matter what the terminating impedance is (including open).

Example: Let frequency be 1 GHz and cable loss be 0.3 dB/meter. Velocity in the transmission line is 66% of the speed of light (c).

Let

$$f \equiv 1 \cdot \text{GHz} \qquad \alpha \equiv \frac{0.3}{20 \cdot \log(e)} \cdot \frac{1}{m} \qquad v \equiv 0.66 \cdot c \qquad \lambda(f) \equiv \frac{v}{f}$$

$$Z_o \equiv 50 \qquad Z_L \equiv 100 \qquad L \equiv 100 \cdot \text{ft} \qquad \lambda(f) = 7.79\,\text{in}$$

With Loss

$$Z_{ic}(L,f) \equiv Z_o \cdot \left[\frac{Z_L + Z_o \cdot \tanh\left[\left(\alpha + j \cdot \frac{2\pi}{\lambda(f)}\right) \cdot L\right]}{Z_o + Z_L \cdot \tanh\left[\left(\alpha + j \cdot \frac{2\pi}{\lambda(f)}\right) \cdot L\right]} \right]$$

$$Z_{ic}(L,f) = 53.466 - 2.374i$$

$$\left|Z_{ic}(L,f)\right| = 53.519$$

$$\arg\left(Z_{ic}(L,f)\right) \div \text{deg} = -2.543 \quad \text{deg}$$

$$\text{Loss} \equiv \frac{0.3}{m} \cdot L = 9.1 \quad \text{dB}$$

Without Losss

$$Z_i(L,f) \equiv Z_o \cdot \left[\frac{Z_L + j \cdot Z_o \cdot \tan\left[\left(\frac{2\pi}{\lambda(f)}\right) \cdot L\right]}{Z_o + j \cdot Z_L \cdot \tan\left[\left(\frac{2\pi}{\lambda(f)}\right) \cdot L\right]} \right]$$

$$Z_i(L,f) = 80.429 - 32.936i$$

$$\left|Z_i(L,f)\right| = 86.911$$

$$\arg\left(Z_i(L,f)\right) \div \text{deg} = -22.27 \quad \text{deg}$$

Appendix D

Measuring Polarization

We learned from the Poincare' sphere that every polarization has a corresponding cross polarization. You can use any pair to create any other polarization. I use the most common way; i.e., "vertical" and "horizontal" polarizations.

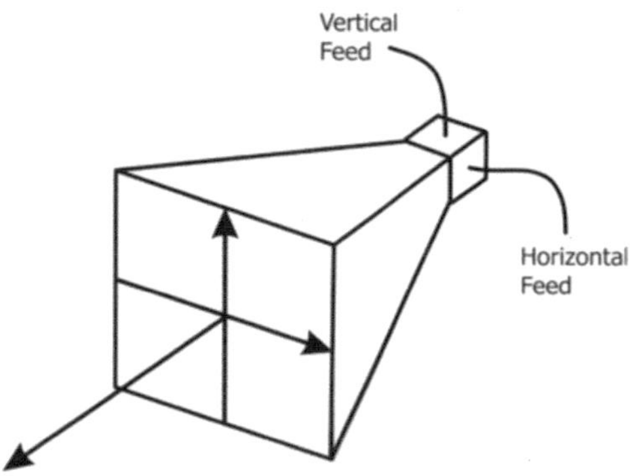

With proper setting of amplitude and delta phase, any polarization can be produced. Common usage is to set linear amplitudes equal and delta phase to 90° to produce circular polarization.

At the receiving end gain adjustment and phase shift can match any incoming polarization.

- As stated previously, **there is only one polarization**
  - The so called Vertical and Horizontal polarization are vector components of that one polarization
- But it is possible to create two orthogonal signals that will sum in the infinite far-field to produce any desired polarization
  - Any two polarizations that are on opposite sides of the Poincare' sphere can be used
    - It usual to use a linear vertical and linear horizontal

Appendix E

Rotating Linear Measurements

Before the days of precision range receivers, the "rotating linear" method was used to measure polarization. As the name suggests, a linearly polarized source (horn) was rotated rapidly. This yielded axial ratio and gain as shown here

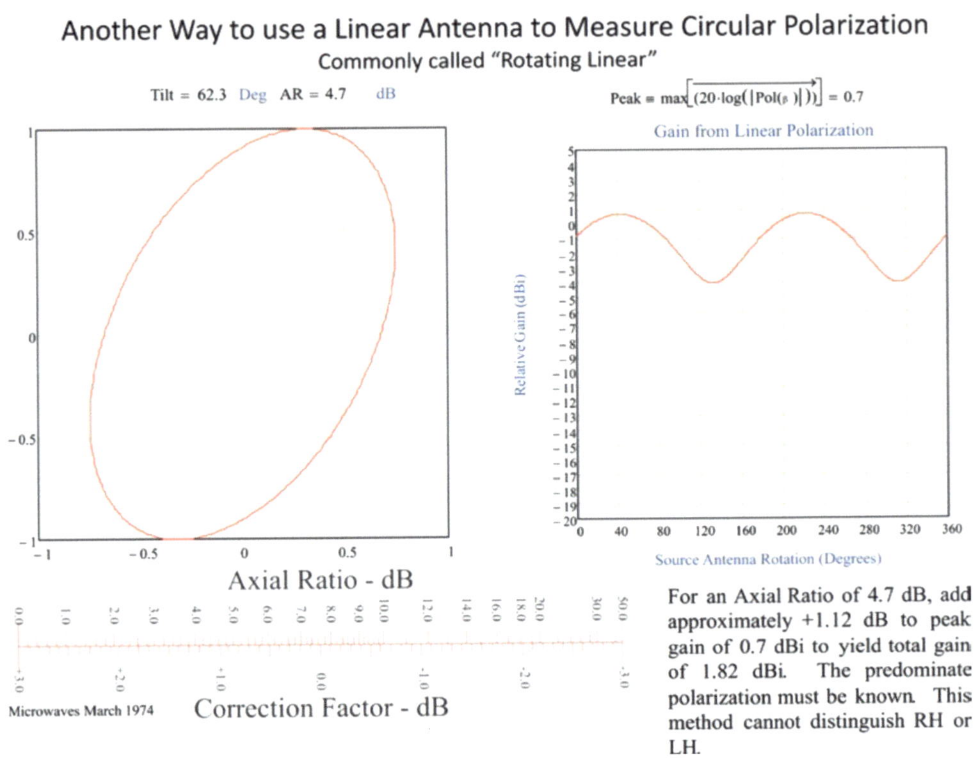

This is a slow and tedious way to measure polarization. It does not lend itself to automation.

Appendix F

Polarization without Phase

To measure polarization without using phase you need a linear source antenna (horn) that can be positioned into four orientations:

Vertical
Horizontal
Slant 45
Slant 135

In addition, you should have a pair of circularly polarized antennas, one RHCP and one LHCP.

Poincare' Polarization Sphere

Modified for Measurement w/o Phase

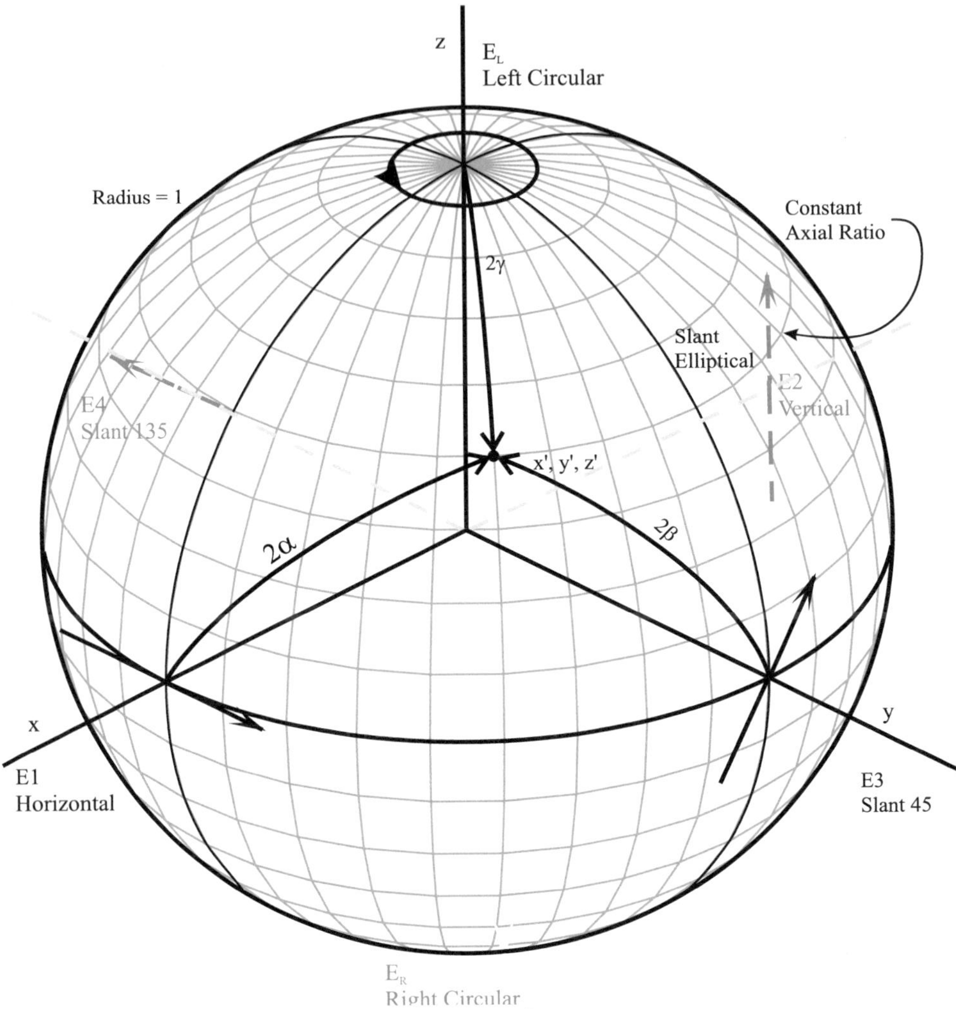

Here are the equations:

Convert Measured Power to Voltage

$$E \equiv 10^{\frac{Ps}{20}}$$

$$\alpha \equiv \operatorname{atan}\left(\frac{E_2}{E_1}\right) \qquad 0 \ge \alpha \ge 90$$

$$\beta \equiv \operatorname{atan}\left(\frac{E_4}{E_3}\right) \qquad 0 \ge \beta \ge 90$$

$$\gamma \equiv \operatorname{atan}\left(\frac{E_R}{E_L}\right) \qquad 0 \ge \gamma \ge 90$$

$$x \equiv \cos(2\alpha)$$

$$y \equiv \cos(2\beta)$$

$$z \equiv \sqrt{1 - x^2 - y^2} \quad z \equiv \cos(2\gamma)$$

(x,y,x should be primed, but my software would not do it.)

Observe that with only the four positions of the linear antenna, we have everything except the sign of z'. In other words, is it left or right polarized? If you know a priori which it is, you can just assign the proper value. Also, if z is very nearly zero, you can just assume it is linear. If neither condition exists, you will need the two circular antennas (one LH and one RH) to make the determination.

Appendix G

Vector Multiplication and Poynting Vector

First let's refresh the fact that the two vectors can be anywhere in space and the following equations are valid. However, the math is easier to understand if we transform them so that the beginning of the vectors are at the origin (0,0,0). Then we can write the vectors as a column of three numbers (x,y,z). There are two ways to take the product of two vectors. Both are extremely important in the solution of Maxwell's equation.

Dot Product

The dot product of vector, V1, and vector, V2 is the magnitude of V1 times the magnitude of V2 times the cosine of the angle between them and it is a scalar. It is also given by (x1∗x2 + y1∗y2 + z1∗z2). Notice at the dot product is fundamental in my solution of the polarization equations. Namely, I use it to find the angle between two points on the Poincare' sphere; which, in turn, yields the polarization mismatch value.

Cross Product

The cross product of vector, V1, and vector, V2 is the magnitude of V1 times the magnitude of V2 times the sine of the angle between them and it is a vector. The vector is perpendicular to the plane containing the two vectors.[58] The direction of the vector is determined by the "right hand rule".

Now let me think "where have I seen this before?". Oh yes, the EM wave is composed of two vectors, E and H, that are perpendicular to

---

[58] Start both vectors at the origin and it is easy to see that there exists a plane containing each of them.

each other and they are travelling in a direction that is perpendicular to both of them!! That feels very much like a cross product.

Now, E has the dimension of Volts/meter and H has the dimension of Amperes/meter. I don/t know about you, but it is hard for me to visualize what they look like in space. But, when we multiply them together we get Watts per square meter, which is easy to visualize. Place an absorber perpendicular to the direction of propagation and see how hot it gets.[59]

Also, W/m^2 is power density which we used to compute how much power a receiving antenna intercepts based on it effective area.

The instantaneous power density is simply the cross product:

$$S = E \times H$$

But we want the average over the period of the EM frequency.

$$\{S\} = \tfrac{1}{2}\,\mathrm{Re}(E_p \times H_p^*)$$

Where Re means "the real part of"[60]
The subscripts mean "peak" value
and * means the "complex conjugate of"
Since the E and H vectors are always 90 degrees apart, the sine of the angle between them is always 1.0.

---

[59] The radiation from the sun at noon time is about 1000 W/m^2.
[60] The Imaginary part (Im) is reactive power that is not radiated.

Here is a graphic to help visualization

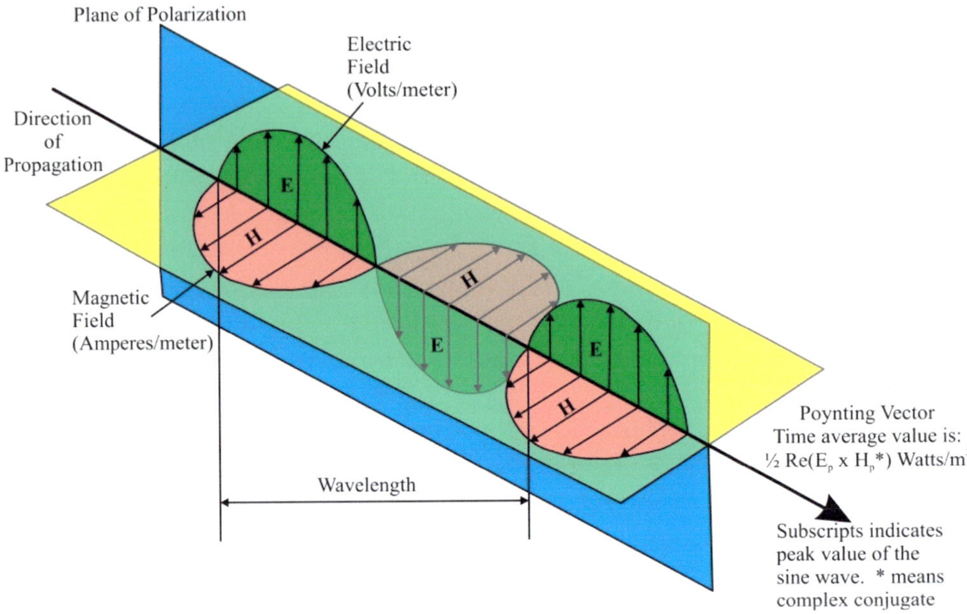

Made in United States
Troutdale, OR
08/30/2025

34103238R00119